THE ICARUS QUESTION

ESSAYS ON SCIENCE,
TECHNOLOGY, & THE
SEARCH FOR HOME
IN A CHANGING
WORLD

GENE TRACY

THE ICARUS QUESTION

ESSAYS ON SCIENCE,
TECHNOLOGY, & THE
SEARCH FOR HOME
IN A CHANGING
WORLD

GENE TRACY

Published by Ramified Press, Williamsburg, Virginia, USA,
ramifiedpress@protonmail.com.
Design by Jason Anscomb, https://jasonanscomb.com
ISBN: 979-8-9882070-0-9 (eBook) | 979-8-9882070-1-6 (paperback)
| 979-8-9882070-2-3 (hardback)
Library of Congress Control Number: 2023908387
Subjects: LCSH: Technology and Civilization | Science — Social Aspects
| Technology — Social Aspects | BISAC: SCI080000 SCIENCE/Essays |
SCI075000 SCIENCE/Philosophy & Social Aspects
| TEC052000 TECHNOLOGY & ENGINEERING/Social Aspects

For Maureen and Kit,
who made it possible.

CONTENTS

Preface

The challenge of finding home in
a world on the move.

We lost our farm when I was seven years old. My parents had tried for nearly two decades to make it work, trying dairy first, then growing corn, followed by sheep and chickens, truck-farming vegetables, and finally setting up an excavation company. But nothing ever really caught on. The local economy in that part of upstate New York appeared to be stuck in a years-long post-war slump while other regions boomed. Disappointed, nearing fifty and without a pension or much in the way of savings, my father went to work as a merchant seaman for over a year, traveling to India, the Mediterranean and North Africa. Meanwhile it was my mother who drove our tan and white VW van down the New York Thruway to our temporary home in New Jersey, where we could be near her sisters and their families.

It was a leaden and overcast November day when we made the journey. We couldn't afford a moving company, and since my brother and I were still too small to do any heavy lifting, my mother had hired a pair of local farmers who were moonlighting during the off-season. We followed behind their rickety flatbed truck, laden with the belongings my parents had decided to keep in the downsizing: a green overstuffed couch, sturdy as a Sherman tank; a rugged, knotty-pine dining room set that my father had made with his own hands; my grandmother's china set; a chest of drawers filled with clothes; several pairs of boots; a few of our toys; two bicycles; a .22 rifle. Peeking out from under the tarp, it made for a tatty collection, threatening to break its ties and spill all over the road. In later years I came to recognize the look from the sitcom *The Beverly Hillbillies* and John Steinbeck's novel *The Grapes of Wrath* (1939): a wandering family, worldly goods piled high, lumbering down the road toward new horizons.

For years I dreamed of returning to the farm. I thought often about my favorite path to the brook behind the barn, no longer open to me now that the brook, the barn, the house and its surrounding hills all belonged to someone else. If my brother and I asked when we were going back, my parents would nod and say, 'one day'. They believed it to be an act of kindness. But we never went back, except for those few times my father tried to collect debts from the men he had thought were his friends. Instead, we moved, and moved again, and so went from owning a hundred-acre farm in New York State to eventually living in a trailer park in Baltimore.

I managed to accept the loss, in time. The seductions of reading, music and writing, and later astronomy—and girls—eventually called me away from pining for the hills and meadows of my childhood. I went to university, where I met my wife and discovered mathematical physics. I made a life as a college professor and raised a family. Yet a lingering sense of not being quite at home in the world has remained—linked, I suspect, to that original, unremedied dislocation.

This book is my attempt to understand how we might make sense of a world that often seems unhomely—one major crisis away from complete collapse, capable of setting us on the road in search of a new way of life and a new home. My own migration experience was in peacetime, and it left our family intact. We moved about within the confines of a modern and stable nation state. The transition was caused not by conflict or environmental breakdown, but because we were part of a wider vanishing of the family farm, and the long, slow decline of the prosperity of rural America. Today much of our world is on the move because of war, political upheaval and forced migration, but most of those people, too, are on the road in search of a better life for themselves and their children. They plant one foot before the other, on a trail lit by the faintest glimmer of hope.

As a scientist, I find myself searching for secular stories of hope. Science combined with storytelling can help us to grapple with existential threats like nuclear weapons and climate change. Yet if science can lay claim to producing reliable knowledge about how the world works, it always draws its strength from what can first seem like a weakness: all good science starts from a position of formal humility, a realization that our understanding is always provisional, and that we must always seek more data. If we stop believing in this potential for surprise, we are no longer scientists.

Because I am a scientist, I believe in the power of theory and observation, and in the fact that rigorous agreement can emerge between the two following open debate and discussion. The philosopher of science Michael Strevens calls such institutionalized doubt, with its demand for empirical evidence for all scientific claims, a kind of 'knowledge machine', and it has, indeed, delivered wonders.[1] But I also believe that a strict scientific approach to the world can never be exhaustive. Making sense of reality is always personal. Love and friendship, fear and hate: these are real and true, and they exist in the world, albeit not in the same way as a mathematical theorem, a planet, or a nucleus. It would seem that those things that make life most meaningful slip past our measuring instruments because they are embodied in flesh and blood, yet they are so much more. We should thank the Universe for the fact that the cold equations of physics have given rise to a warm world, redolent with the potential for empathy and compassion.

The ancient Greek tale of Icarus and his father, the inventor Daedalus, can speak to us here. They, too, were dislocated, held captive against their will and tried to flee to a better life. Like us, in their efforts to solve one problem they created others. Trying to escape by using artificial wings, Icarus ignored his father's plea and flew too close to the Sun, melting the wax in his wings and plunging him from a great height into the sea.

Like Icarus, we might also be undone by our striving nature. Do we believe that science and technology can make the world a better place, or that they inevitably lead to new catastrophes? Can we succeed in our interlinked quests for new knowledge, for ultimate power, and for material wealth and control of the planet? Will this urge to transcend our human limitations, to follow our curiosity and our desire to shine the light of our intellect into all the dark places—will it necessarily lead to disaster, or will we always find a way out of any predicament?

These essays are an offering in response to these questions. I have attempted to explore multiple aspects of the rapidly shifting ground of our technologically mediated modern cultures. These cultures can be deeply alienating, even for those of us who grew up as technophiles immersed in some version of them. That makes it all the more important to seek out glimmers of hope, for the ways that we might create a home within the world we are making. Otherwise, we risk becoming rootless, setting out upon the road without a good internal map of our outer physical and social realities. We risk becoming isolated individuals awash in a sea of data, our societies atomized into human archipelagoes. This would be tragic, because we are all eventually carried by the passage of time into a form of exile from the world of our youth. If we are lucky enough to grow old, we are destined to be shipwrecked on some foreign shore called the future, surrounded by strangers that we can only hope will show us some kindness.

As we try to imagine what the future might hold, the question that should concern us is not whether Icarus can avoid disaster, but can he learn to swim?

Introduction

In order to thrive in the coming decades, we must not only strive to better understand the world, but we must also love it more passionately.

I once polled my cosmology class and asked them: how many years does our species have ahead of us? Ten thousand? A million? I was stunned by the pessimistic tone of the discussion that followed. The most outspoken students talked in terms of hundreds of years, while the others simply sat there in glum silence. I later ran across one of the quiet students and asked her: why weren't there at least a few hopeful voices willing to speak up that day? She replied that she did believe we had a long future ahead of us, but she couldn't articulate why, and felt unable to contribute. The pessimists seemed so sure of themselves, almost smug in their certitude that doom lay just around the corner.

Dark stories permeate pop culture: looming existential threats, nuclear war, climate catastrophes, plagues, societal collapse. Those prone to despair have many cultural references to draw upon, from the nightly news to science fiction dystopias in film and literature. I realized that I'd failed my young optimist, and students like her, by not giving them some conceptual tools to defend a more balanced outlook. This book is my overdue response to the pessimists, and one scientist's argument that we must try to create a climate for hope. We can do this by designing for surprise, by remaining open to new possibilities.

Hope is not the same as benighted optimism. Bald optimism and simplistic pessimism are both downright perverse. Both are forms of denial and excuses for inaction, and their shared certainty betrays a lack of humility. They are retreats from a sincere engagement with the question: what steps must we take right now to increase the odds of human survival and flourishing in the long-term?

Though nothing is guaranteed, this book lays out a case that we can still choose an open-ended future. The optimist needs to acknowledge the chance of failure, while the pessimist must be open to the possibility of success. While it won't be easy to get past this century's compounding crises, we might yet create a more humane society, one that is also far more benign in its effects on the rest of the living world. But it will be hard work.

This leads us to the heart of an old problem. Humans have had a fraught relationship with our inventions from the very beginning. Consider the stories from Greek mythology of Daedalus the inventor, and his son Icarus. Icarus is the far more familiar figure in popular culture today, a symbol of hubris, youthful ambition, or the urge for transcendence. Using wings fashioned by his father and held together with wax, Icarus receives a stern dad-lecture about not flying too high. But the young boy ignores the warning and seeks to fly close to the Sun to meet the god Apollo in his fiery chariot.

Less well known today is the figure of Daedalus, though he could stake a claim to the status of the greatest inventor of all time in the tales of the ancient world. He supposedly invented the saw, the compass, and many other tools that made further inventions and machines possible. This was many centuries before the flowering of Ancient Athens, when power in the eastern Mediterranean radiated outward from the Court of King Minos of Crete. The reputation of Daedalus catches the attention of King Minos, who recognizes the potential for technology to increase his own power and wealth. And so, Minos brings Daedalus, and Icarus, to serve in his court at Knossos. Daedalus realizes that by placing himself in service to such a powerful man, he can gain access to extraordinary resources to carry on his work.

So, Daedalus builds Talos, a giant mechanical guardian in the shape of a human which circles the island of Crete three times a day. When it encounters an enemy of King Minos, it chases them down, hugs them to its breast and roasts them alive. An even stranger story of the inventiveness of Daedalus concerns a request by Queen Pasiphaë, Minos' wife. She's fallen in love with a bull and convinces Daedalus to create a kind of robotic shell for her in the shape of a cow. This allows her to crawl inside and sidle out among the other cows in the pasture, attracting the bull so it will mate with her. This bizarre tryst leads to the birth of a beast that is half-human, half-bull: the Minotaur, which must be confined and hidden away.

Daedalus, ever helpful and handy with designs, creates a giant Maze beneath the palace to confine the monster, to which Minos then feeds his human sacrifices. By now Daedalus starts to wonder if this is, perhaps, not the best situation in which to raise a young boy. So, he begins to plan for their escape from the high tower in which King Minos has imprisoned them both. But how? By watching the birds soaring freely outside his prison window, he hits upon a scheme.

While the subplot of Icarus is often read as a warning against aspiring to godhood, the story of Daedalus is more of a cautionary tale about the unintended consequences of our inventions. Humanity is like Icarus, always striving to get above ourselves, but we are also like clever Daedalus, in that we sometimes unleash tragedy upon an unsuspecting world.

Thus, we are led to The Icarus Question: can he learn to swim? The answers we might give reveal something deep about our relationship to technology, our faith in human ingenuity, and our intuition about the future.

Though it might be tempting to do so, we must first resist thinking there will be one single future, shared by all. The past is messy, so there's every reason to think the future will be, too. Therefore, we must do the creative work of envisioning a menagerie of futures, a disordered heap of possibilities, and so create a world in which each of us can choose where we would feel most at home. Some might prefer to live in cities, others in rural areas closer to nature. Some might wish to live on the Moon. Why not? We need to develop an intuition for that tendency of complex adaptive systems to fission into new forms and shapes. Giving others permission to make lives they find meaningful implies that we, too, have a right to share in the joy.

Second, designing for surprise means that we need to adopt an attitude of humility about our current state of knowledge. That doesn't mean we don't know enough to act on things like climate change. We do. But making wise choices right now means acting in a way that enlarges the field of play for future generations, to give wider scope to the human future. Mistakes are inevitable, but we can learn, change and adapt as science advances, and as we see how our current choices play out.

Take the climate predictions for 2100, which involve large uncertainties. Those outcomes depend strongly upon actions we might take or fail to take. Designing for surprise means giving our children and grandchildren the gift of greater options from which they can choose when the time comes. We need to embrace our innate ability to reinvent ourselves, our institutions, and our forms of artistic expression, if we are to navigate our way to a better future. At the same time, we must acknowledge that all human creations are brilliant in conception, flawed in design, and botched in execution. Hence the need for humility, and the inclusion of exit ramps on the road to any given future. Designing for surprise means that in a world of accelerating change we must learn to become ever more nimble.

Humans can be pig-headed and cruel, but we are also capable of great compassion and generosity, and of intellectual feats of nuance and subtlety. We begin our lives as citizens of the world, roly-poly cosmopolitans, infants able to learn any living language. We quickly become bound to a specific culture through a sort of forgetting, a closing off of possibilities. But we do not then become frozen, unable to change and grow.

If current demographic trends continue, most current college students will still be alive in the year 2100. This telescoping lifespan, paired with the increasing pace of change, means that intellectual and creative neoteny—the extension of youthful and even childlike characteristics well into adulthood—will be ever more vital.

Humans can be like the tricksters of myth, a shape-shifting animal with no fixed nature. The Coyote figure of Navajo tradition is credited with causing a primordial flood, interfering with the placement of the stars, and generally poking his nose into everybody's business to mix things up. The Raven that features in the creation stories of the Haida, who traditionally live off the coast of what is now British Columbia in Canada, may have simply been attracted by a bright and shining light, as ravens are, but when he stole the Sun from its hiding place, he brought light to the whole world.

These characters are neither good nor bad, but agents of chaos that don't always know their own power or the likely outcome of their actions. By cleverness and guile, they continually set the world on a new path, and thereby keep it from getting stuck or running down to stasis. Like them, if we remain open to amazement and novelty, we can also become something new, even in our old age.

Techno-optimists point out that science can deliver innovative cures for disease, reduce the likelihood of famine, and increase human lifespans. These are sources of hope, of course, and they reflect the practical benefits of science. But science also uncovers the remarkable beauty in the world around us, as it is now, and ideally shares its discoveries freely with everyone.

These essays partly spring from my journey as a scientist who has managed to remain curious, continually astonished by the beauty of our world, as well as the kindness of strangers. These experiences are my primary sources of hope. Science is not a haphazard collection of random facts, an ever-expanding cabinet of wonders, but it instead advances through the use of systematic methods in the pursuit of generalizable knowledge. It diligently seeks out connections and patterns. Like skeins of knitting that form a growing whole, scientists see knowledge as a web of interconnected probabilities, built upon personal experience and study, and the legacy of the generations that have gone before us. The scientific mindset of practical humility, generosity of spirit, and openness to wonder and surprise can also be a powerful source of hope. These are values that science shares with the wider culture when both are at their best.

Science is not alone in pursuing conversation across generations. Enabled by a constellation of technological and social innovations such as writing, the printing press and the university, this potential for learning across generations is perhaps the most important social invention in human history. And yet this effort of millennia appears to be coming unglued before our eyes. Books and longer forms of writing can seem too slow, struggling to keep up with the rapid pace of change and the shifting grounds of meaning. Many of us are moving from cultures where knowledge is viewed as largely longitudinal in time to one that is more lateral, where we seek wisdom in the self-absorbed moment, not in the ruminating ages. This is true not only of the young but of all generations now caught up in these tectonic social movements. The shift in our internal map of knowledge can lead to the seductive trap of the eternal present, and it carries momentous implications.

These essays are about how we adapt to change, and how we might maintain the strengths of past practices even with altered tools and times. They explore the ways we learn and grow as individuals and societies, using the methods of literary non-fiction to look at how science and culture must co-evolve in the face of civilizational crisis. A major thesis beating beneath the essays is that in order to thrive in the coming decades, we must not only improve our scientific understanding of how the world works, but we must also use our heart to love the world more passionately. I mean not only our love of the natural world and other human beings, but also of human invention and creativity. Because if you don't love the world, why would you bother to understand how it works, and begin to make things better? And if you can't imagine a better world, why would you even try?

CHAPTER 1

The Mind Shaper
and the Maze Maker

*Stories shape us and allow us to find a way to the
heart of what we are searching for.*

❧

'It is a paradox of the labyrinth that its center
appears to be the way to freedom.'
—MICHAEL AYRTON

According to legend, everyone in the Marshall Islands descends
from two goddesses, Liwātuonmour and Lidepdepju, both of
whom arrived long ago in a canoe from the west. After a time,
they turned to stone: Liwātuonmour on the island of Namo, and
Lidepdepju on the island of Aur. To the eye of later colonizers,
the resulting structures looked like coral outcroppings, not unlike

many others in that part of the world. But to a believer, they were gods frozen in motion. Thereafter, because of the belief that greater powers resided there, people on Namo and Aur did not have to bow before the Irooj (Chief).

In the late 19th century, Dr C. F. Rife, a Baptist missionary from America, sailed to Namo and removed Liwātuonmour from her resting place. The islanders told him that if he moved the stone, the goddess would destroy his boat. Dr Rife carried her out beyond the reef where he tossed her overboard in deep water, apparently to demonstrate to the startled islanders that it was just a rock after all.[1]

Old gods die hard. For many centuries after Christianity came to Northern Europe, folk legends survived that the ancient pagan gods were not false, but instead had been defeated by the Christian God and driven into exile where they now lay in wait, brooding on revenge. From centuries of experience, the Christian missionaries knew that to defeat the old gods you must run them to ground: to find their source of power and kill them where they hide, deep in the labyrinth of the human mind.

What was it like for the Marshall Islanders present that day, men and women who witnessed their goddess thrown into the sea by a white man? Blasphemy. And then: nothing. It must have seemed a world-shattering event when Dr Rife turfed Liwātuonmour out of the boat. Yet the sun came up the next day, the bananas ripened, the islands still rode above the waves, and the monsoon rains still fell from the skies. How terrible this must have been, this normalcy on the heels of spiritual outrage.

Those who have suffered a crisis of faith know it can strike like a bolt into the core of your being, turning your soul to ash. Yet these are also the moments when old myths get rewritten, when the world becomes new again, when our mental landscape undergoes a tectonic shift. That act of rewriting after trauma is necessary for survival; otherwise, we can remain as if trapped in that moment, unable to move on.

This is a point made by Jonathan Lear in his book *Radical Hope: Ethics in the Face of Cultural Devastation* (2006). In that book, he relates the story of Plenty Coups, a Great Chief of the Crow Nation who lived during the years when his people were driven onto reservations and forced to abandon their nomadic lifestyle. Close to his death in 1932, in speaking to an interviewer, Plenty Coups said that once his people were confined, unable to practice the old culture, 'nothing happened'. It's tempting to interpret this as meaning that nothing happened that was worth creating new stories about.

If correct, this reflects the urgent need to use storytelling to move past trauma. The 2018 movie *Black Panther* brought the cultural movement known as 'Afro-futurism' into mainstream popular culture, but the title character first appeared in 1966 as a Marvel comic book hero, created by Stan Lee. As literary and cultural movements, 'Afro-futurism' and its cousin 'indigenous futurism' cover themes such as recovery after loss, renewal, and revenge, or explore parallel histories where European colonialism didn't prevail.

The term 'indigenous futurism' was coined by literary scholar and author Grace Dillon, who also edited *Walking the Clouds: An Anthology of Indigenous Science Fiction* (2012), the first collection of indigenous futurism stories ever published. Using the literary tropes of science fiction, these authors imagine possible futures where native peoples continue to play a large role in world history, settling Mars, undoing past wrongs, or parallel histories where Māori warriors sailed up the Thames in an ocean-going canoe and sacked London. Telling these stories is part of the struggle for cultural survival and renewal in what—for indigenous peoples—is a post-apocalyptic world following the arrival of European colonizers.

Such stories do not reflect a rejection of the modern world, but instead represent an assertion of agency, a right to take part in imagining what the future might hold for everyone. They depict an abundance of possible futures, inhabited by all peoples, and the world-building reflects an even wilder abundance of cultural referents—going far beyond the European-inflected civilizations in *Star Trek* or *Star Wars*.

———

Writing and reading are late additions to our mental toolkit, invented long after the evolving human mind became adept at storytelling. It's not surprising, then, that most ancient stories and legends exist in multiple forms. The tribal band that huddled round a campfire was an early focus group, and itinerant poets would have sensed which way their audience leaned, who their listeners wanted to live, and who they believed should die.

Consider the Greek story of Iphigenia. She was the daughter of Agamemnon, and falsely promised by him to the warrior Achilles. Iphigenia's ambitious father and her false lover Achilles (who was in on the lie) bound the young girl in her wedding robes and brought her to an altar—not as bride but as a blood sacrifice to ensure fair winds for Agamemnon's fleet of ships, to carry his men and arms to war with Troy.

There are multiple versions of what happens next. In the *Oresteia* by Aeschylus, Iphigenia dies on the altar. Agamemnon has his eventual victory at Troy only to be murdered by his wife upon his return, leading to a bloody cycle of revenge and the eventual destruction of the House of Atreus. In another telling, the goddess Artemis takes pity on the young Iphigenia and turns her into a doe that springs from the altar and escapes the sacrifice at the last minute. In yet another ending, the gods spirit Iphigenia away from the altar to live with them, forever young and beloved.

There are also various tellings of the story of Icarus and his father, Daedalus—maze maker, master inventor and gifted technician, who gave himself over in service to the ambitious King Minos of Crete, and eventually tried to flee with his son. In the most familiar version of the story, Icarus dies when he ignores his father's warning and flies too close to the Sun, causing the wings fashioned by his father to break apart. In another, they escape from the tower by other means, then attempt to sail away from Crete. But Icarus allows himself to become distracted and falls out of the boat and drowns.

In each telling, Icarus dies as a result of his own youthful self-regard, but what changes are the meanings of his death. While Iphigenia is innocent, and therefore worthy of pity, the tragedy of Icarus is that he must die as a victim not only of his own failings—his arrogance and self-absorption—but also of his father's ambition. Icarus dies because of his urge to transcend his human limitations, to touch the blazing sun of freedom itself.

﹌﹋

My friend Al Osborne is familiar with this deeply human impulse. He grew up in Houston in the 1960s, a heady time for a boy who, like me, was a fan of the space program and dreamed of becoming an astronaut. With the 1961 founding of what was later called the Johnson Space Center, Houston quickly became the headquarters for human spaceflight in the US.

Upon hearing that NASA was building a new facility nearby to support the Apollo program, Al set out in his car to the site and found a single operator driving a bulldozer in an open field. He approached the man and said: 'I want to help y'all get to the Moon. Where do I sign up?'

Al was directed to a construction trailer nearby. From there, knocking on doors and following leads, he eventually arranged an internship at NASA while he was in college. That led to a job after graduation helping to put men on the Moon, working every day with the Apollo astronauts in training.

The metaphor of the open field, of open lands more generally, or unexplored seas, where all things seem possible: humans itch to be at play in that unfenced area, building new dreams, looking over the next hill. A similar image must have imprinted itself upon us thousands of generations ago, when our ancestors in Africa decided, for reasons we might never know, to spread throughout the continent—eventually striking out across the Sinai Peninsula and into the world beyond, becoming global humanity as we know it. However much we change in coming generations, we owe it to those ancestors to keep that restless, roving part of our heritage alive.

Yet the urge to wander can be counterproductive, and we might wonder why evolution didn't breed the instinct out of us. Moving into new lands, encountering new plants, new animals and new predators, new terrains and weather patterns—all these increase the threat of an already dangerous world. But while the urge to wander can be dangerous for the individual, it's also true that it's perilous for any species to get too comfortable in their ecological niche, which can change remarkably quickly. All species must hedge their genetic bets, and the urge to explore is probably vital for the long-term survival of the species.

With that epochal step out of Africa, humanity began the long wandering that brings us to the present day. We now fill every available niche on the planet, including outposts at the South Pole and in near-Earth orbit. Our Daedalus drive pushes us to look in all directions, not only over the next hill, but into the very heart of the atom, the workings of light and matter, and to understand the secrets of life itself. Hence, we are brought to the current state of things, where as a species we are bending the previously undirected processes of the Earth toward objects of the human will—inadvertently changing the chemistry of the atmosphere and oceans in ways that harm not only ourselves, but the entire biosphere that sustains us. Ninety-five percent of the biomass of large non-human animals is now domesticated, and those domesticated animals are treated as commodities, suffering in a system of economic exploitation. So, we are driven to seek a new land just over the mental horizon, a place more in tune with the natural world, and more humane. The better angels of our novelty-seeking natures know there must be a better way to do things.

Even the simplest of human actions has complex and unexpected consequences. To follow the Daedalus drive to expand our knowledge of the perimeter, to better understand the center, to explore space and the atom—these efforts require skills and technologies that, unfortunately, can also be turned to war and conquest.

As a young German, Werner von Braun started an amateur rocket club in the early 1930s. His youthful obsession with space exploration, carried into adulthood in a time of war, led to his willingness to work for a genocidal regime building weapons of mass terror. Exploiting people interred in concentration camps, he became the man who built V-2 rockets for the Nazis to attack London, and then Saturn Vs to help the Americans beat the Soviets to the Moon.

The fatal binding of science, technology and war has been one of my enduring preoccupations. Like the deadly dance between Daedalus and King Minos, scientists use and are used by the nation states in which we live and enjoy rights of citizenship. Many of us have skated on moral thin ice at various times in our careers. Many are also patriots, but we tend to be a globalized and cosmopolitan elite as well—one that, unusually, still enjoys relatively high public approval ratings. Compared to politicians, bankers, and the media, scientists and engineers are seen as heroes.

Nearly all scientists will say they work in the public interest, and most believe that they do. Yet our research can unleash mayhem, not necessarily with ill intent, but because we are usually more concerned with chasing down a technical question than with worrying about its wider social impacts. Many years ago, when I proposed the creation of a course about the physics of nuclear weapons, I was visited by one of the senior faculty in my department and given a stern lecture about not mixing science and politics. He called on me to disassociate my moral self from my intellectual self, to sustain a bifurcated state of being in order to achieve the necessary emotional distance the scientific mindset supposedly demanded. I did not always find this easy.

Often these tensions play out on the plane of geopolitics. In the summer of 1990, the Congress of People's Deputies met in the Kremlin to debate whether to dissolve the Soviet Union. I was attending a mathematics conference at the same time, hosted by the Joint Institute for Nuclear Research in Dubna, a large research facility in Russia, roughly equivalent to the Los Alamos National Laboratory in the US. Conference attendees were ensconced in an isolated compound. Sleeping quarters for guests were spartan by Western standards, but more or less on par with my room in Moscow at the 'luxury' Hotel Intourist. At meals we ate good borscht and caviar, drank good wine and vodka, and chatted amiably about science and mathematics in a wood-paneled dining hall.

After a famous Soviet scientist regaled me about the difficulties he was experiencing due to the latest political upheavals, I remarked that at least he was privileged as a mathematical physicist. Hard as his life might be at the moment, surely the average person and the poor had it much harder. How would those people get by if the system collapsed? He simply shrugged his shoulders, as if he found the question uninteresting. Perhaps I'm being unfair; perhaps he was simply overwhelmed and had grown numb.

We talked about something else. Meanwhile, outside the science compound, the Soviet Union was in free-fall. I saw depressed and desperate drunkards stumble across on-ramps to highways and piss on the dilapidated walls of the local post office, unremarked by passersby.

≫⋅⋅≪

Some of us seem addicted to remaking ourselves from time to time, adopting the latest technologies and living on what's called the 'bleeding edge'. Such people crave new things, perhaps finding that life has no salt unless our whole world view is upended from time to time. Perhaps these periodic attempts at reinventing our inner lives are a substitute for a bygone ability to physically wander over the next hill. As a result of all this exploration and invention, science has taught the rest of us the truth of things that were once unthinkable: *We are not the center of the universe. We are an animal that shares a common ancestry with the apes.* Some scientists even claim that it's possible now to answer age-old questions once considered within the exclusive remit of philosophy or religion: *There is no soul, only biochemical networks. Our precious and unique self ends when the body dies.* These are old ideas, a century or so old at this point. But new speculations, sometimes posing as verities, seem to have arrived: *The day when intelligent machines will transcend us is fast approaching. The pace of change is accelerating, and soon things will be changing too fast for human cognition to keep up. We may be living in a simulation.* And, finally, the dawning realization that, *by our own hand, we have set in motion one of the world's great extinction events.*

The mind recoils at these ideas and seeks a way back to safety in the familiar.

So, in a sense, we are always living in an Icarus moment. In writing about science and technology, we might be tempted to invoke wonder and astonishment, curiosity and surprise; we speak about exploring the unknown, expanding our horizons and the length of our reach. But one person's astonishment is another's existential angst. Some willingly implant technology in their very bodies, while others squirm at the thought of it. We tell stories and write poems about Icarus, but no stories are told, no songs sung, about those who hold back, those who—paralyzed by a fear of flying or perhaps merely an abundance of caution—think Icarus mad for trying to touch the blazing heart of the world. But by denying the existence of a response to pull back, to not risk everything for a chance of ascending to a kind of godhood, we ignore ancient wisdom. That caution, after all, is what kept our human ancestors alive down through the eons long before we became the top predator. A tendency to dismiss caution as cowardice or moral failure makes it all the easier for the Daedalus drive to double down and carry on.

And so, we drift toward chaos, where many societies seem to be bifurcating in myriad ways—not merely into Red States and Blue States, Right and Left, liberal and illiberal regimes, but between those who believe Daedalus must follow his drive wherever it leads, and those who believe he and his followers will eventually be destroyed by their overweening ambitions. The divide is between those who think Icarus can somehow learn how to swim, and those who believe he must inevitably die.

≈⤙⤚≈

Tossing a goddess out of the boat, as the missionary Dr Rife did to the Marshall Islanders, seems a ready metaphor for what the Daedalus drive in science has done to our understanding of ourselves. It can make societies throw away their moral touchstones. Such a process can numb us, like repeated blows. We must make our minds strong, but what about our hearts? The survival of the species depends upon whether we can pursue our Daedalus impulse without killing ourselves; whether we can shine light into all the hidden corners of the cosmos and return with cures and new astonishments; whether we set foot upon all the solid worlds of the solar system—and all the while, whether we can avoid ever more mechanized slaughters, successors of the killing fields and the genocides. These are the ultimate outrages against human wellbeing.

The story of Daedalus is about a man co-opted by a king whom he tries to co-opt in turn. That move and countermove are all too familiar. While Daedalus strives to create new wonders, his work also leads to the fashioning of a monster such as the Minotaur—one that had to be hidden away, trapped in a maze below ground, feeding on the flesh of young men and women. The centuries between the rise of the Minoan civilization and the fall of the wider Hellenic world over a thousand years later saw true advances in science, technology, metalworking, and textiles. But very little of it was turned to making life better for the average man or woman, many of whom were slaves. Few of the benefits of advancing science were turned to labor-saving inventions or machines of mass production, to extending the general human life span and prospects. Almost all the benefits flowed to a select few at the top of the social pyramid.

Humans now living in the Global North are the inheritors and beneficiaries of a cultural legacy that emerged quite recently in historical terms—one that fetishizes innovation and change as having value in and of themselves, an invented tradition that applauds creative destruction as a permanent and desirable state of being. As a scientist, I once enthusiastically embraced this scheme of things. Our society's support for science and technology justified the creation and taxpayer support of institutions that nurtured people like me—physically awkward but book smart, with a facility for numbers. Someone, in other words, who would not have lasted long in a hunter-gatherer society. Now, at the end of my career, my enthusiasm for scientific discovery endures, but it's tempered by a sense that we are not doing enough to make sure the discoveries of science and technology are shared widely. I worry that we are not doing enough to allay the anxieties of those outside the sciences, for whom things seem to be moving too fast toward some unknowable and dark future. We scientists need to attend more to helping people find a home in this new world that science has uncovered.

Science, at its best, relates to the unknown with what philosophers sometimes call both 'epistemic hunger' and 'epistemic humility', meaning that we should always approach the world with open-eyed wonderment, full of curiosity and a desire to learn new things. We should overflow with questions but treat all answers as provisional. The remarkable advances in science can sometimes cause researchers to become hubristic, to claim that science is the one true path to wisdom or even worse, to assert that questions that can't be answered empirically are simply meaningless.

Such arrogance can slip easily into the totalizing and sterile ideology of *scientism*. At times it can seem as if the reshaping of the human mind by advancing scientific knowledge requires the smashing of whole armies of idols. But creative destruction can provoke reactionary countermoves. For example, some scientists seek a 'god gene' that would explain why some people are more prone to mystical experience than others. Are they Dr Rifes in a lab coat, now thumping a biochemistry textbook rather than a bible? I have scant ability to judge the quality of this research, or to interpret the evidence. But how the research is talked about in public discourse matters. Nuanced language is called for, and a respect for human values which one might not share. Otherwise, a fear of cultural annihilation can, in turn, risk a rejection of science entirely, which would be tragic.

In the 1940s and 1950s, the Marshall Islands were the site of 67 above-ground nuclear tests. The plunk of the goddess into the waves of Aur and the mushroom clouds of Enewetak and Bikini are separated by only fifty years and a few hundred miles. They both reflect human drives to overcome. In the 19th century, the Marshall Islanders gave up their old gods and warlike ways, embracing a Christianity that was strongly urged upon them. The Islanders also voluntarily vacated Bikini and Enewetak a few generations later, in answer to a call to help the US build a weapon so terrible that it would bring an end to war itself. An Icarus impulse, a striving to liberate ourselves from vulnerability.

If we revisit the story of Icarus, we might now imagine Daedalus, father of Icarus, wings broken by his own hard landing, as he steps to the edge of a rocky cliff to overlook the Aegean. The image of the boy's plunge is still fresh in his memory. Exhausted and grieving, he spies nothing but empty waves lit by the setting sun. He falls to his knees. Bloodied by the rocky ground, he gathers a handful of the thin soil and pours it over his bowed head in silent lamentation.

All his days and nights, all his scheming to bend the wealth and power of King Minos to his own advantage, to build machines born of his own imagination and to shine the light of his own hungry and questioning mind into all the dark corners of the world—those efforts have led to both wonders and monsters. During the escape, wings strapped to their shoulders and ready to ascend, he had counselled humility to Icarus. But the boy was too much like his father.

Now the Sun sinks low. Far out to sea, the lengthening blood-red shadows are broken by a single hint of white. Perhaps the foam of a breaking wave. Or something more—a wing? Does it move? The mind of Daedalus says it cannot be so, but his heart yearns to believe. Has Icarus learned to swim?

Before we can ponder the answer to the Icarus Question—before we decide whether he can learn to swim—we might ask: are we still imprisoned in King Minos' tower, about to take flight, and with time to reconsider? Or are we like Daedalus, alone and bereft, having seen our beloved child plunge from a great height, a victim of hubris? If our feet are still earthbound, then the tragic heart of the myth concerns our forward-looking instinct; it warns us that Icarus will die if we launch ourselves into the air. If so, we should back away. If instead, we see ourselves near the end of the myth, with tragedy behind us, Icarus having vanished from sight into the waves below—then perhaps we must believe Icarus will swim, or else we must give up all hope and descend into the quietus of despair.

Yet we can tell more hopeful stories, too, where invention is used to make the lives of people better. Another story from the Marshall Islands tells of a woman named Lōktanūr, who had many sons. One day, the sons decided to build canoes and have a race. They would paddle across the lagoon to another island and the first one across would be chief. As they lined up their canoes on the beach, their mother appeared carrying a bundle on her shoulder. She went to her eldest and said: 'Take me aboard.' He refused, saying she would only slow him down. She turned to the next oldest, who also refused to let her ride with him. Lōktanūr went down the line of sons until she finally got to her youngest, Jebro. She told him: 'It doesn't matter if you are chief, for you are the youngest. Take me in your boat.' He agreed. With that, Lōktanūr unfurled what she had been carrying in the bundle: a mast and a sail. The other brothers had already begun the race, so Jebro and his mother started out far behind. But in the end they won easily.[2]

Thus, the Marshallese Daedalus was a woman, and her invention opened the wider world to her people. As a result, her sons, especially the faithful Jebro, prospered. As we ponder these

stories, Daedalus and Icarus, Lōktanūr and Jebro, we should remember that these narratives unfold in the elastic realm of mythical time. We might imagine Jebro and his mother happily at sail,
beating eternally against the wind. Or the reach of Icarus toward
the blazing heart of things, and that plunge from a great height,
might take a day, a generation, or an age. But it could also seem
like a fleeting moment, especially when viewed from a distant future that knows how our story ends.

CHAPTER 2:

Children of Apollo

*Rockets are dreams made into fuel and fire. And if we
forget to dream, or if we dream only other
dreams, we will lose the sky.*

⟞⟞⟝⟝

Some of the fondest memories from my American childhood are
of watching rocket launches on television. First came the Mercury spacecraft that put humans into orbit, which was followed
by the Gemini program that proved it was possible to walk in
space, as well as to carry out orbital rendezvous and to dock and
undock. Then the Apollo series brought it all together, putting
humans on the Moon and returning them to Earth. Alan Shepard
and John Glenn, and later Neil Armstrong and Buzz Aldrin, were
all household names in my family.

It all happened with blazing speed, and then vanished like a
meteor from the nation's imagination. As Marina Benjamin relates in her wonderful memoir *Rocket Dreams* (2003), today it's
hard to grasp just how rapidly it all took place—how driven we
were as a nation to set foot on the Moon, how uncertain it was
that we would reach that goal, and how quickly we turned away

when it was over. The recent film *Apollo 11* (2019), based on ar-
chival footage and narrated in the words of those who took part
in its events, captures that sense of energy and collective purpose.

On live television, the drama played out for all the world to
see. In 1962, US President John F Kennedy declared the goal of
landing a man on the moon by the end of the decade. In July
1969, less than seven years later, Armstrong took a first step onto
the lunar surface. At its peak NASA absorbed over 4% of the US
Federal Budget, and most of that went to human spaceflight. The
pace of felt time quickened, but it soon slowed once more. By the
mid-1970s the Apollo program was a shell of its former self, and
its last two missions were canceled. A further mission to Mars
was also scrapped, as was the construction of a space station that
would be the site of a massive project orbiting the Earth to support construction of the Mars spacecraft. The current International Space Station (ISS) is a less ambitious version of that original design. Of that earlier, grander vision, only the Space Shuttle
remained, like a high-priced bus system with no place to go.

All this tumult filled my teenage brain. At the age of twelve,
watching Armstrong's small step while living in a trailer park in
Baltimore, I vowed to myself that I would one day walk on the
Moon. Like those participating in the human space program,
I would become a scientist, and then an astronaut. America
was still the land of opportunity; we were heading into space,
and quickly. I became a fan of science fiction and dreamt of exploring other worlds, of meeting alien races. My grandmother,
concerned for my mental health, tapped me on the shoulder
one day while I was watching *Star Trek*, and asked me to assure
her that I didn't really believe all that stuff. How could I explain
to her, someone born before the automobile, who had lived
through two World Wars and the Great Depression—how could
I explain that my future was going to be different from her past,
that we were on the verge of something new, that the state that

lay before us was one of permanent astonishment?

At the time, in the late 1960s and early 1970s, all astronauts were military pilots with academic pedigrees. So, when an Air Force recruiter came to my high school, I asked about signing up. I was a painfully shy and skinny working-class kid: the classic geek, long-haired, socially awkward and not athletic, with poor eye-hand coordination, dreamy and forgetful. And asthmatic. The recruiter took one look at me and managed not to laugh, though I detected a smirk. On that day, I realized I would have to give up my dream of becoming a pilot and flying to the stars. But I was good with numbers and science, so if I couldn't fly space-ships, at least I could try and build them.

My high school was intended for vocational students going directly into work upon graduation, not for those aspiring to attend college. Our only guidance counselor told me outright he couldn't help me decide where to apply, or how. I was lucky that I tested well, and that US society at the time wasn't overly concerned about credentials and 'extracurriculars'. High school students weren't expected to have impressive CVs before even applying to college. So, I had a shot, and John Hopkins took a chance on me, for which I will always be grateful.

Though Hopkins was only a half-dozen miles as the crow flies from where I lived, it might as well have been on another planet. The college catalogs in my high school library were woefully out of date, and the campus visit had not yet become part of the American rite of passage. When I finally arrived, thinking I would join the aerospace engineering program, I was shocked to discover the department had shut down. The tides of Apollo funding had ebbed just as quickly as they had flowed only a few years before. I have a vivid memory, possibly concocted by my overactive imagination or a dream, of standing before a small and derelict red-brick outbuilding, ivy covering its walls and windows, with a pigeon cooing from its roost in the dormer window overhead. On a small wooden sign hung by the door on a rusting nail, the cracked paint read: *Aerospace Engineering*. So that dream became dust as well.

Clearly my game plan had to change once again. I decided to shift from engineering into physics, and then to learn celestial mechanics. If I couldn't fly or even build spaceships, I could at least tell them where to go. As it turned out, the shift in goals from flying in outer space, to building spaceships for others to fly, to the more abstract world of theoretical physics, ultimately had its compensations. It was there I fell in love with mathematics and its idealized models of the world. Theoretical physics was a good place to follow the siren call of beauty while hiding for a spell from the real, messy world, one that seemed to be coming apart in dangerous ways. When I got the chance to do a PhD in mathematical physics at the University of Maryland, I jumped at it, and never looked back.

❦

Fast forward thirty years to Wallops Island, Virginia. While I've long been a fan of space exploration, I had never seen a launch in person until September 2013. That day, I traveled with my child, Kit, to the Mid-Atlantic Regional Spaceport (MARS) to watch the launch of a space probe known as the Lunar Atmosphere and Dust Environment Explorer (LADEE). I was familiar with Wallops Island as a site for suborbital flights, those that reach space but whose trajectory intersects again with the Earth's atmosphere, so they don't achieve orbit. Over 16,000 have been launched from the facility there in its seventy-year history, but Orbital Sciences, who was responsible for LADEE and is one of a new breed of private launch companies, aspired to something more ambitious: to send a probe to the Moon. To top it off, it would be a night launch.

The evening was bathed in mild late-summer weather, with

few clouds. We found a crowd gathering in Chincoteague, Virginia, about five miles from the launch site. Lift-off was scheduled for around 11 PM. We arrived a few hours early and began spreading out our lawn chairs. Families and their children ate picnic dinners around a large-screen display NASA had set up. Right on time, the crowd hushed for the final countdown.

Far in the distance, to the south, what looked like a sudden sunrise lit up the horizon. A small bright flame leapt from the ground, rising higher and higher, until it moved among the stars themselves. Then the ground rumbled as the sound finally arrived. A young boy near us shouted 'Dad! It's on fire!' When his father didn't respond, the boy shouted again: 'But why is it on fire?'

How do you explain how a rocket works to a small child? But then, how do you explain the meaning of it to an adult who has lost the capacity for wonderment? How do you explain to a student learning the Newtonian mechanics of rocket propulsion that the explanation you have just given them is correct, but also misses the point? The mathematics, while beautiful, are desiccated and bloodless, cold equations devoid of some crucial sense of why a rocket strikes a spark to the tinder of the imagination. A rocket is a tool of mythical power because it can take a piece of the everyday human world and fling it to the stars. That's why a night launch is so poignant, because the tail of the rocket isn't overwhelmed by the Sun's glare or the blazing blue sky; instead, it becomes one with the night and takes part in its awesome splendor.

The author Kurt Vonnegut witnessed one of the Apollo launches and called it a kind of orgasm, an experience which unsettled the usually unflappable and sardonic man. The Orbital Sciences LADEE rocket I saw at Wallops was not nearly as big as Apollo's Saturn V. Anyway, phallic symbolism seems to me far too small-minded and petty to capture the phenomenon we'd witnessed. The true power of it can't be captured in ideas, stories, images or metaphors. To feel it in your bones you must see it first-hand. Rockets are dreams made into fuel and fire. And if we forget to dream, or if we dream only other dreams, we will lose the sky.

When the LADEE spacecraft joined the silent stars, I found myself moved. Watching it chase the Moon, I was reminded how much I yearned to go myself.

Now nearing the end of my career, I find my mental and physical wanderlust remain. Hence my ongoing interest in space exploration, my curiosity about the endless depths of space, the workings of the atom, and the austere beauty of mathematical theorems. I feel lucky that this is so, because it keeps my mind young while my body ages. Yet I've often wondered where this fascination in things beyond the horizon comes from. It's hard even for scientists, or perhaps especially for scientists, to speak in rational terms about what drives them. I think this is because ultimately what moves us is not rational. It is primal.

During the first Bush Administration in the US, around the time when Congress was debating whether to fund the International Space Station (ISS), D. Allen Bromley, the president's science advisor, came to visit the university where I taught. At a meeting with the physics faculty, I mentioned something that was worrying me: given the tight budgets in the space program, and cutbacks in science funding more generally, should we be building a space station that had no clear scientific objective? The Mars and Moon programs had been canceled years earlier, so the Space Station wasn't even being touted as a waystation to anywhere. Critics complained it would be an expensive pressurized tin-can in low-Earth orbit, and a made-up destination to create a purpose for another troubled program, the Space Shuttle. Major science organizations had come out publicly against the ISS because of fears that cost overruns would gradually eat into the NASA budget for robotic exploration of deep space, and other important research projects like the Hubble Space Telescope (HST), which revolutionized our understanding of Black Holes and star formation.

Because the NASA budget must cover both crewed and robotic missions, the tension between human spaceflight and uncrewed exploration at NASA continues unabated. The Artemis program, which aims to return humans to the Moon, is experiencing cost overruns and is far behind schedule, yet it is managed by the same agency and covered by the same overall budget as projects such as the James Webb Space Telescope (JWST). This superb telescope is now on station a million miles from Earth. Able to detect longer wavelengths than the HST and with greater spatial resolution, the JWST can see the earliest light emitted by galaxies formed shortly after the Big Bang. It is already adding to our understanding of phenomena ranging from Near-Earth Asteroids that might one day be a threat to Earth, to the search for the chemical signatures of life on exoplanets that orbit other stars. All the various landers and deep space probes that NASA, the ESA, and other space agencies have launched over the years have widened and deepened our understanding of the Universe. They have delivered a sustained and ongoing revolution in astronomy and astrophysics, and they didn't require a human presence to do so. Hence my question to Bromley all those years ago: given the finite resources available to NASA and the other space agencies, and the success of robotic probes, why did he believe it was important to send human crews out into space?

Bromley answered by arguing that the earlier Apollo program, and later the Space Shuttle and the ISS, were our 'cathedrals'. For a nuclear physicist to use the rhetoric of the sacred among his peers astonished me. But where we reach for our metaphors often reveals the wellspring of our true motivations. Cathedrals, too, were not enterprises motivated by reason. They were physical manifestations of spiritual aspirations for transcendence—vaulted spaces to promote wonder and awe, to foster a sense of God's presence in the world, to guide our vision upward toward heaven. In the current age, the aspiration to surpass human limitations has, for some, become a desire to quite literally ascend into the heavens and create a multiplanetary civilization. So perhaps it's true that dreams of cathedrals and spaceships have the same point of origin in the human heart, nestled there alongside our need for hope in a better, more open-ended human future.

Today the pace of felt time seems to be quickening once again. There is renewed planning for a return to the Moon, of asteroid prospecting, of settlement on Mars. Even so, building a spacefaring civilization will be the work of centuries. Before we can terraform Mars, we will have to navigate the compounding crises we face on Earth. We stand on the edge of forever, and our choices matter because they can shift the long-term odds in our favor, or against us. If we give up on our home, we risk losing our only chance at the stars.

CHAPTER 3:

Sky Readers

For most of human history, from ancient wanderers of the Eurasian Steppe to the Polynesian voyagers, the stars told us where we were in space and time. Have we forgotten how to look up?

❦

Some years ago, I visited a gallery that specialized in Inuit art, and the owner shared with me a small but powerful memory. She had a close relationship with many of her artists, one of whom had shown up for a visit late on a winter's night. Like many of us, the first thing he did was call home on his cellphone, to let his wife know he had arrived safely. Unlike most of us, especially on a frigid winter's night, he did so out in the yard. He needed to see the sky so he could tell his wife what the stars looked like from Richmond, Virginia, while she scanned the sky at her end, far away in Hudson Bay. In that way, he connected with her, by finding one another through that useful intelligence of distant stars.

For most of human history, this artist's behavior would have seemed ordinary, even essential. It was unthinkable to ignore the stars. They were critical signposts, as prominent and useful as local hills, paths or wells. The gathering-up of stars into constellations imbued with mythological meaning allowed people to remember the sky; knowledge that might save their lives one night and guide them home. Lore of the sky bound communities together. On otherwise trackless seas and deserts, the familiar stars would also serve as valued friends.

> 'When all the stars were ready to be placed in the sky,
> First Woman said, "I will write the laws that are to govern
> mankind for all time. These laws cannot be written on the
> water as that is always changing its form, nor can they be
> written in the sand as the wind would soon erase them,
> but if they are written in the stars they can be read and
> remembered forever." —Navajo creation story, as quoted
> in George Johnson's *Fire in the Mind: Science, Faith and
> the Search for Order* (1995)

That friendship is now broken. Most of us don't orient to our loved ones using the lights in the sky, nor do we spend our nights pondering what in the 1920s the poet Robinson Jeffers called that 'useless intelligence of far stars'.[1] Discoveries in astronomy and physics of the past century have expanded the known universe by orders of magnitude in size and age and turned cosmology into a true observational science. Those breakthroughs urged upon us an extraordinary stretch of the imagination, even as related technological advances detached nearly everyone from that larger world, relegating the stars to the realm of instruments and making those distant orbs safe for us to ignore.

Today we are more disconnected from the stars than ever before. Even utilitarian attachments have fallen away, as the markers that form our sense of place in the wider world have shifted from the distant to the local. Navigators once used the stars as reference marks; the GPS units in modern cell phones refer instead to a constellation of artificial satellites in orbit around the Earth, synchronized to atomic clocks in ground-based laboratories.[2] (There has been one intriguing reversal of the trend: anxiety about the wartime vulnerability of the GPS system recently prompted the US Naval Academy to reinstate the teaching of celestial navigation.[3] This particular unease is an apt metaphor for our general anxiety about losing our way when the lights go out, and about where we stand in relation to the world.)

In untethering ourselves from the stars, we have lost a part of ourselves. Knowing where you are in the world is fundamental to knowing who you are. The development of our sense of spatial relationships—the ongoing discovery of where *I am*—is deeply entwined with the formation of memories. Neuroscience suggests that this is because developing the knowledge of place and building that sense of our relation to other parts of the world, requires the brain to combine several different sense modalities. Hence information must be stored and then retrieved from memory, sifted and examined, and the brain's theory of where we are in the world constructed. Combine this with the fact that it is through memory that the *I* endures, that memories are most effectively formed when there is some emotional charge attached, and we see why our sense of place can be so entangled with our sense of who we are—and why to be at *no place* is akin to being *no one*.[4]

In addition to a resurgent interest in navigating with the stars, many scientists, engineers and visionaries are now engaged in imagining how we might navigate *to* the stars, not only as a question of technological hurdles to be overcome, but also as a point of aspiration to keep humans aimed toward the unknown. As an example of this literature from the last generation, I am drawn to *Interstellar Migration and the Human Experience* (1986),[5] an edited anthology about our species' possible out-migration toward the stars. It is a curious mix of articles, careful scholarship leavened with wild ideas. The contributors argue that, rather than a wholesale assault on the outer solar system or a colossal Apollo-style long-jump to Alpha Centauri, a more likely model for some future wave of human expansion is a long, anarchic diffusion.

Over generations, the human home would expand to include the outer moons of the Solar System, the minor planets, the cometary debris of the Kuiper Belt, and then the distant Oort Cloud, whose outermost objects are likely to wander between star systems. The book's vision of the future makes our move to the stars look a lot like the Polynesian Diaspora, which also took many generations and involved hopping from one solitary foothold to the next across miles of empty ocean.

The Polynesian Triangle stretches from Hawaii in the north, to Rapa Nui (Easter Island) in the southeast, and Aotearoa (New Zealand) in the southwest. This region of the Pacific is comparable in area to the entire Eurasian landmass. When Europeans 'discovered' these islands, they found most of them already inhabited. The Polynesians all had similar languages and customs and, despite the great distances between the islands, they maintained contact with one another. At Raiatea (near Tahiti) in 1769, the English explorer Captain James Cook took onboard a Polynesian navigator named Tupaia who drew a map for him showing islands within roughly 1,000-mile radius of Tahiti. How did Tupaia know this, Cook wondered? How could he keep that map in his head?[6]

The navigators of Polynesia (who in traditional societies were men) found the ocean and sky an open book, full of information. Traditionally trained navigators had a memory map containing hundreds of stars. If the navigator could get a clear shot of an open patch where the stars shone through, even if much of the sky was covered in clouds, he could orient himself and maintain a heading at night.[7]

Navigational lore in the Pacific Islands was passed down through the generations partly through storytelling. The human mind is structured around stories. Connecting things to stories, poems, songs, music and visual art makes this knowledge more real to us, charged with emotive power, which aids in the forming of memories. It helps us come to know things, and to know their place, by knowing ourselves more deeply as well.

An anthropologist would rightly point out that there is no bright line distinguishing a testable theory of the world from a story. They are not the same thing, but neither are they completely different. They are alike in that theories and stories are both ways of organizing knowledge and making sense of things. In the case of Polynesian voyaging, experimental knowledge that was part-ly encoded in stories allowed navigators to find their way across hundreds of miles of open ocean. This knowledge was replicated over many generations, tested every time a crew set out into the open Pacific, and verified every time a boat made landfall.

By the time of Johannes Kepler and Galileo Galilei, when 17th-century European astronomers finally began to shake off the lingering dreams of Aristotelian cosmology, the Polynesians had been abroad for more than two millennia. The mental maps of the Pacific Islanders were organized not around Aristotle's crystalline spheres of quintessence, but around poems, songs, and tales of skittish parrotfish. The Polynesians were crossing thousands of miles of open Pacific Ocean while the Greeks were still hugging the shores of the much smaller Mediterranean, composing epic poems to celebrate their own intrepid voyages.

In the Western tradition, the overthrow of the Aristotelian worldview by Copernicus and his followers involved more than moving the Earth from the center: eventually it called for the abandonment of the potency of place, a central tenet of Aristotle's thought. Aristotle believed it was self-evident that there was a special place, and that everything in the world sought to find its proper location in relation to that center. In contrast, modern cosmologists often invoke what has come to be called the Copernican Principle: we are at no special location in the Universe. Rather, we are wanderers among the stars, through a space with no center.

To a modern cosmologist, then, there is no special place. Yet Aristotle was onto an important psychological truth: our sense of place is indeed potent. The vast deep that the astronomers have uncovered all around us—and yet so far have found to be empty of conversation—that ringing silence is the source of our fascination with science fiction based on other worlds and stories of aliens. We don't want those newfound playgrounds for the imagination to be empty. We want to find ourselves again among the stars. We want those stellar nurseries to be full of squalling civilizations. We want them to be other *places*, other centers of meaning, not just beautiful imagery. We want the cosmos to be full of stories, so we can imagine coming to know it.

While Polynesian navigators were internalizing stories to help them find their way at sea, Western astronomers gradually came to realize that, moving from a direct visual experience of a celestial body to an isolated entry in a logbook, each step introduces possibilities for error, misinterpretation, personal bias and confusion. Hence, as Western astronomy became more complex—as it evolved into a science in the modern sense—its practitioners worked hard to reduce the human role to more machine-like actions.

They emphasized repetitive, routinized steps. In order to minimize bias, the human beings involved in the daily work of the 19th-century's great European observatories were told no more than they needed to know to carry out their simple tasks. This made the observatories more like a factory than the romantic vision we have of a devoted individual, driven by love of the stars, carrying out a solitary vigil on a mountain top.[8]

At Greenwich Observatory in the mid-19th century, for example, the routine jobs were carried out by dozens of boys in their late teens or early 20s, each specialized to make a simple observation and to call them out, to mark the time, or to reduce data. It is no wonder they were called 'drudges'. They did the menial work while the foreman of the shop floor scanned the skies—not directly, but as part of a superorganism, gazing out at the stellar vault with the perspective of the Astronomer Royal, George Biddell Airy. The collective activity buzzed around him at night while he studied the motions of the stars and planets, all that effort bent toward generating the tables of ephemerides needed by mariners for celestial navigation. That was crucial intelligence for the world's preeminent global empire, with its fleets of merchants and dreadnought battleships, all indexed to the far stars. Those pinpoints of light remained unmoving throughout the years, forming a fixed frame of reference for finding oneself in the world—the same sky text that the Polynesians had used when sailing the open Pacific, and that the Inuit artist used to phone home.

Nearly all the routine human operations of the astronomical past are now carried out by electronic or mechanical analogs. The alienation of the drudges and human computers is almost complete. They've been replaced by servomechanisms, avalanche photodiodes and computers, while the results of querying the heavens are digitized and shared through cloud applications. The artificial senses used by astronomers and physicists have revealed that the world around us is stranger than any earlier generation had ever imagined. Yet these revelations can't be reduced to mere collections of data. Technology is simply a tool that can open a new window. What we *see* while peering through the window, how we absorb it into our internal sense of things, how it shifts our sense of our place in the world, that fuller act of *seeing with new eyes* requires a lively imagination.

~~~~~

As a physics undergraduate, I once attended a seminar by a graduate student reporting his observations of one of the first stellar black hole candidates, an X-ray source known as Hercules X-1 (the first X-ray source discovered in 1971, in the constellation Hercules). His few minutes of X-ray data were collected by a detector he had built for himself and installed atop a suborbital rocket. He then launched the rocket from a former oil platform off the coast of Kenya. Other than that romantic locale, what I remember most about the talk was a question from the audience: 'If I went out tonight and looked up, where would I find Hercules X-1?'

The student was flummoxed. He didn't know. His life those past months had been consumed by getting his detector to work, arranging for delivery of the payload to Kenya and then shipping it to the San Marco platform near Malindi; then he had to get it properly installed aboard the rocket, as well as endure the tension of the launch and the wait to find out if he had collected the data he needed for his PhD thesis. If he failed, he would have to start all over again. In all those months, he had not looked up to the stars, even for curiosity's sake, to see with his own human eyes the region of the sky where he would point his X-ray eyes.

As I proceeded further in my studies, I came to understand how this could happen. In my efforts to study the stars as an undergraduate, I found I had no time to ponder them. I eventually turned aside from the narrow path one followed to become a professional astronomer, and instead became a mathematical physicist. But I never lost my love of the stars, and I still enjoy teaching about them.

Part of the alienation of the astronomer from her objects of study comes down to the remarkable technology that allows remote, automated observation. This technology stands between us and the stars as surely as a window at once separates us from nature while opening it to view. Unlike those who can take a walk in the woods to be close to the things they love, the astronomer can only go and hug a telescope. The beauty of far Antares, the red star at the heart of the constellation Scorpius, provokes a love forever unrequited. This kind of alienation, formed by peering through newly opened windows at things forever out of reach, is particularly acute for the working astronomer, but it affects all of us who attempt to navigate the virtual world that our interactive screens have brought into view.

It is not just the stars we have learned to ignore. How many of us remember phone numbers anymore, or email and street addresses? Our memory is becoming ever more externalized, stored on the cloud somewhere—we don't know where—nor do we care so long as we can access it when we need it. While the information seems mentally nearby, the stored memory might physically reside on a server 1,000 miles away. We are ever more augmented human beings, ever more virtualized, enjoying what the philosophers Andy Clark and David Chalmers call the 'extended mind'.[9]

How is the extended mind changing us, our sense of our selves, and our sense of our place in the world? How does it affect our sense of being at home in the Universe, or our general level of unease, given we don't know where many of our memories are stored, nor how many physical copies there are and how safe they might be against erasure? And if those essential parts of our lives are treated so blithely—those wedding photos, those names and addresses of friends and loved ones, those journal and diary entries—if those can be remembered 'out there', are we at risk of gradually assuming that our personal story can also be virtualized and become part of our extended selves, detached from any particular place? If to be at *no place* means you are somehow *no one*, doesn't this progressive externalizing of memory increase the risk of our story fading into nowhere?

In 1959, the Italian physicist Giuseppe Cocconi and his American colleague Philip Morrison published one of the first articles to examine the question of how humans might listen for interstellar signals, thereby launching what has come to be called the Search for Extraterrestrial Intelligence (SETI). Morrison later wrote about the commonality between our urge to explore other planets, the search for conversation among the stars, and the role of storytelling among hunter-gatherers of Africa's Kalahari, a people who spent their lives in a region of space no larger than Los Angeles County. While they wander, they constantly discuss where they are, so as 'not to fade off into the nothing or the nowhere'. He argued that 'this is the essential feature of human exploration, its root cause deep in our minds and in our cultures'.[10]

In his essay collection *The Dream of Spaceflight* (2000), the space evangelist Wyn Wachhorst claimed that we explore the periphery so that we can complete the center. We range widely so that we can come home changed, more fully realized, knowing better who we really are. We are like wolves restlessly prowling the limits of our territory, orienting ourselves, discovering our story by finding our place in the world, and a place for those we love. I like this metaphor of exploring the periphery and the related need for completion, because it emphasizes why we all need stories, even in this techno-civilization we are creating. We need stories to undo the post-Copernican flattening of things, the externalizing of self, so we can live more like the Kalahari hunter-gatherers who notice when something has changed in their world—and not like those harried professionals who walk, oblivious, past astonishing wonders, caught up in the everyday sameness of the synthetic world.

This marking out of place in order to bring it within the realm of story is the impulse behind the recent—possibly temporary—

naming of mountains and frozen plains on Pluto, names such as
'Cthulhu' and 'Balrog'.[11] Pluto, ancient god of the underworld,
now has dimples and warts on his face named after more recent
mythical creatures of the deeps, born of the 20th-century imaginations of H. P. Lovecraft and J. R. R. Tolkien. The image analysts
for the *New Horizons* spacecraft had grown tired of speaking to
one another of 'that dark patch on the left side of the computer
screen', so they did the same thing our ancestors did when they
named the stars and gathered them into constellations. These
21st-century explorers connected features to stories to make
them easier to remember, to organize, to fire the imagination, to
guide understanding. To help extend our sense of place, our sense
of home, even on the outskirts of the solar system.

If humans do someday migrate outward toward the stars, our
narrative space will move like an expanding wave before us, a
vanguard of the imagination. Our need for stories that help us
find our way is too important to be left behind.

As a small child growing up on a farm, I wandered the place
with my brother, two intrepid explorers of far pastures and deep
woods. I wanted to know every tree, to look every cow in the eye
and come to know its thoughts. I wanted to turn over every rock
in the brook and find the lurking crayfish, knowing that every
single moment of time was charged with a magic that is now
largely gone from my experience of the world—an electric arc of
being that I now glimpse only occasionally when I glance up at
the night sky and my mind is at rest.

I understand what Morrison meant when he wrote about the
need for our stories 'not to fade off into the nothing or the nowhere'. For me, it connects to a desire to rekindle my earliest sense
of wonder, my sense of place and who I am, my sense of being at
home and at one with the stars.

CHAPTER 4:

Learning to See

One astronomer's dimpled pie is another's cratered moon. How can our mind's eye learn to see the new and unexpected?

When Galileo looked at the Moon through his new telescope in early 1610, he immediately grasped that the shifting patterns of light and dark were caused by the changing angle of the Sun's rays on a rough surface. He described something akin to mountain ranges 'ablaze with the splendor of his beams', their sides in shadow like 'the hollows of the Earth'; he also rendered these observations in a series of masterful drawings.[1]

Six months before, the English astronomer Thomas Harriot had also turned the viewfinder of his telescope towards the Moon. But where Galileo saw a new world to explore, Harriot's sketch from July 1609 suggests that he saw a dimpled cow pie. Why was Galileo's mind so receptive to what lay before his eyes, while Harriot's vision deserves its mere footnote in history?[2]

Learning to see is not an innate gift; it is an iterative process, always in flux and constituted by the culture in which we find ourselves, as well as the tools we have to hand. Harriot's six-power telescope certainly didn't provide him with the level of detail of Galileo's 20-power. Yet the historian Samuel Y. Edgerton[3] has argued that Harriot's initial (and literal) lack of vision had more to do with his ignorance of chiaroscuro—a technique from the visual arts first brought to full development by Italian artists in the late 15th century.

By Galileo's time, the Florentines were masters of perspective, using shapes and shadings on a two-dimensional canvas to evoke three-dimensional bodies in space. Galileo was a friend to artists, and in his youth might have considered becoming one himself.[4] He believed with a kind of religious fervor that the creator of the world was a geometer. Galileo likely imbibed these mathematically deep methods of representation, based as they are on the projective geometries of light rays. Harriot, on the other hand, lived in England, where general knowledge of these representational techniques hadn't yet arrived. The first book on the mathematics of perspective in English—*The Art of Shadows*, by John Wells—appeared only in 1635. When Galileo looked at the face of the Moon, he had no trouble understanding that there, mountaintops first catch fire with the rising Sun while their lower slopes remain in darkness, just as they do on Earth. Galileo, therefore, had a theory for what he was seeing when those pinpricks of light winked into existence along the line of day and night; he even used the effect to measure the heights of those lunar mountains, finding them higher than the Alps. Harriot, a brilliant polymath yet seemingly blind to this geometry, looked at the same scenes half a year before Galileo, but didn't understand.

When we consider scientific observations—those paragons of a purportedly objective gaze—we find, in fact, that they are often complex, contingent, and distributed phenomena, much like human vision itself. Assemblies of high-powered machines that detect the otherwise undetectable, from gravitational waves

in the remotest cosmos to the minute signals produced by spinning nuclei within human cells, rely on many forms of 'sight' that
are neither simple nor unitary. By exploring vision as a metaphor
for scientific observation, and scientific observation as a kind of
seeing, we might ask: How does prior knowledge about the world
affect what we observe? If prior patterns are essential for making
sense of things, how can we avoid falling into well-worn channels
of perception? And most importantly, how can we learn to see in
genuinely new ways?

〜〜

Scientific objectivity is the achievement of a shared perspective.
It requires what the historian of science Lorraine Daston[5] and
her colleagues call 'idealization': the creation of some simplified
essence or model of what is to be seen, such as the dendrite in
neuroscience, the leaf of a species of plant in botany, or the tun-
ing-fork diagram of galaxies in astronomy. Even today, scientific
textbooks often use drawings rather than photographs to illus-
trate categories for students, because individual examples are
almost always idiosyncratic; too large, or too small, or not of a
typical coloration. The world is profligate in its variability, and the
development of stable scientific categories requires much of that
visual richness to be simplified and tamed.

   In 1890, the meteorologists Hugo Hildebrandsson, Wladimir
Köppen, and Georg von Neumayer published a most unusual attempt at idealization. Theirs was the first cloud atlas, proposing to
standardize terminologies and categories of wisps of water vapor.
To earlier generations, this had seemed a hopeless project. Clouds
are things of infinite variety and shape, a drifting canvas for day-
dreams. But the creation of agreed-upon names and categories—
the feathery high-altitude cirrus, the brooding low-lying stratus,
the puff-ball cumulus—this parsing of the visual world overhead

and the creation of a shared vocabulary was a great advance of 19th-century meteorology.

Clouds, as it turned out, are a helpful proxy; they allow large numbers of semiskilled observers on the ground to visualize the weather conditions overhead. Following the simple instructions provided by Hildebrandsson and his colleagues, which eventually became an international scientific effort, the observers would know how to report observations of cloud types and the directions of motion, along with the local weather. Taken over the course of a year, commencing in May 1896, this compendium of information would allow meteorologists to puzzle out the pattern of winds in the upper atmosphere for the first time.

I was interested to see the *International Cloud Atlas* (1896) first-hand. Even though I am a theoretical physicist, I still harbor a need to hold something in my hands, and to see it with my own eyes, before I believe I've grasped its nature. Although the Atlas has been updated many times, and the current version is available online,[6] there are not many copies of the 1896 Atlas left, so I had to take a road trip to visit the Special Collections Library at the University of Virginia. Instead of the book I expected, the librarian presented me with what looked like an artist's portfolio, bound by a ribbon: eight pages of text and a collection of 28 photochromotypes that were as fragile as old family photographs.

Photographers had only recently learned how to capture the sky in a manner that brought out the shapes of clouds, and so the first Atlas is a mix of photographs and artful renderings in paint and pastel by human hands. The foregrounds of the chromotypes are particularly intriguing as attempts to frame and contextualize forms that would otherwise be free-floating. A water tank pokes into the sky in the photograph of the cirrus; rooflines in Paris are photographed beneath mottled stratocumulus; a young man dawdles on the bank of a river, in pastel, gazing up at low-lying stratus; and a sailing ship is painted upon a calm ocean, eternally under way beneath a hazy bank of altostratus. The need to supplement the photographs with drawings and paintings is an example of idealization at work. The creation of objective categories of cloud-shapes thus involved a collaboration between art and science, driven by a love of shape and form, and the uncovering of an orderly beauty in the sky overhead.

In our urge to find patterns, we are like a rock climber pulling ourselves up the sheer wall of the world. We drive a piton in when we find a handhold of pattern, a small crevice of meaning, some slight imperfection in the rock face. As a physicist, I believe that this works only because there is already some opening there to grab on to. To say that we construct idealized categories is not to say that patterns in the world don't already exist, but that we must learn how to see them in the world around us.

~~~

Hold up your hand in front of your face. How can you see what's there? Parsing meaning from randomness—the signal from the noise—is fundamental to both sight and scientific observation. Unless we are blind, our open eyes are flooded with photons at every moment, a noisy stream of information that is then launched from the retina, traveling as electrochemical impulses along the optic-nerve pathways. These are taken up by neural assemblies and, in the dark cavern of the skull (filled to the brim as it is with a sloshing assembly of soft matter), the brain sifts that welter of data for signals that conform to particular patterns. (Fingers? Check. Five of them? Check.) Some neural assemblies focus on detecting certain shapes, such as edges or corners; others specialize in collecting those shapes into higher-order schemes, such as a coffee cup, the face of a friend, or your hand.

These internal visual elements are a mix of predilections that we are born with, and patterns learned from personal experience;

how they affect our perception varies according to our under-standing and expectations. When the midcentury psychologists Jerome S. Bruner and Leo Postman presented test subjects with brief views of playing cards, including some non-standard vari-eties—such as a red two of spades, or a black ace of diamonds—many people never called out the incongruities. They reported that they felt uneasy for some reason but often couldn't identify why, even though the reason was literally right before their eyes.[7]

So, crucially, some understanding of the expected signal usually exists prior to its detection. To be able to see, we must know what it is we're looking for, and predict its appearance, which in turn influ-ences the visual experience itself. The process of perception is thus a bit like a Cubist painting, a jumble of personal visual archetypes that the brain enlists from moment to moment to anticipate what our eyes are presenting to us, thereby elaborating a sort of visual theory. Without these patterns we are lost, adrift on a sea of chaos, with an unsettling sense that we don't know what we are looking at. Yet armed with such forms, we risk seeing only the familiar. How do we learn to see something that is truly new and unexpected?

Vision is not only personal and patterned, but also complex and spatially distributed. In the 1970s, Elizabeth K. Warrington and Angela M. Taylor studied patients with brain damage in the posterior regions of the brain, but with no apparent damage to their visual pathways.[8] It transpired that the part of the brain that's active when we identify the three-dimensional shape of an object (say, a cylindrical white item on the desk) is different from the area involved in knowing its purpose or name (a coffee cup that holds your next sip). Warrington and Taylor tested subjects' performance on a 'same shape' task by presenting photographs of familiar objects, such as a telephone, but taken from both stand-ard and nonstandard perspectives.

For the 'same function' task, all the objects were photographed in a standard view, but subjects had to group them according to function—for example, to distinguish different sorts of telephones from a mailbox. People with lesions in the right posterior region showed no deficit when objects were presented 'normally,' but displayed a reduced ability to identify objects when they were seen from unusual perspectives. Roughly speaking, the results were consistent with the subjects being unable to rotate the object in the mind's eye into a more 'standard' orientation for comparison with internal idealized mental models. Meanwhile, people with lesions in the left posterior brain showed little deficit in such geometric categorization yet displayed a reduced ability to identify the object's name or function.

Subsequent research such as functional MRI imaging on healthy patients has led to a more nuanced understanding of these matters, but the key insight remains: our extraction of a visual signal is a process distributed across the brain, a silent chorus of the mind that, when it works, effortlessly generates visual meaning.

If the brain is a taxonomizing engine, anxious to map the things and people we experience onto familiar categories, then true learning must always be disorienting. Learning shifts the internal constellation of the firings of our nerves, the star by which we set our course, the spark of thought itself. This mental flexibility is an undeserved inheritance, hard-won over eons by our ancestors, and it serves as a good metaphor for how scientists can learn to see with new machine eyes.

～～

In science, seeing things afresh sometimes demands a concerted (and contested) shift in paradigms, such as the move from Ptolemy's map of the planets to those of Copernicus and Galileo. On other occasions, it happens by accident. All the output of our instruments is signal, in a fundamental sense; noise is just that part we are not interested in. Separating out the signal, then, depends upon who is doing the observing, and for what purpose. The 'noise' picked up by Arno Penzias and Robert Wilson in 1964 in their microwave detectors at Bell Labs in New Jersey turned out to be a clue to one of the most astonishing discoveries in science. After heroic efforts to reduce the background hiss detected by their instrument, they determined that it was not coming from the instrument itself, nor from its suburban surroundings. Instead, they became convinced that it was omnidirectional, and coming from the sky. Penzias and Wilson eventually realized that the hiss was part of the cosmic background radiation, a remnant 'signal' from the Big Bang. Now, 50 years later, the detailed angular distribution across the sky of statistical fluctuations in that radiation is a key piece of evidence that leads many cosmologists to believe in dark energy.[9]

The scientific projects that best stretch our understanding of what it means to see in new ways, and reveal the distributed nature of observation, are those that span the globe and involve thousands of individuals, such as the Laser Interferometer Gravitational-Wave Observatory (LIGO). Efforts such as LIGO are often described as 'opening a new window' on the cosmos—yet no human being sees a gravitational wave with LIGO, just as no single neuron in our brain perceives a bluebell flower. In both cases, seeing is a multipartite process, requiring a comparison between noisy signals and idealized models.[10]

More than a century ago, Albert Einstein predicted that a warping of space and time propagates at the speed of light, like the outgoing ripple from the snap of a rug. These space-time ripples are gravitational waves. Consider an incoming burst of gravitational waves, emitted from a distant galaxy by a colliding pair of black holes. Gravity is extraordinarily weak compared with the electromagnetic forces that hold most objects together in our world, giving everyday stuff a solidity that we take for granted.

The weakness of gravitational waves therefore means that they have completely negligible effects locally, unless we isolate massive objects from one another and then allow them to move freely in response to the passing space-time ripples. The gravitational waves generate oscillations in the relative positions of these massive objects, not unlike the motions of leaves afloat on the surface of a pond as wavelets pass among them.

If Einstein was right, we are bathed in these ripples, but we have been blind to the information they carry—until now. In the case of LIGO, the passage of a gravitational wave produces a shudder—a fraction of a nuclear diameter—in the relative positions of two pairs of 40-kilogram mirrors, four kilometers apart. Splitting a specially prepared laser beam in two, bouncing it off these mirrors, and allowing the two laser beams to interfere again, the LIGO detector measures a time-varying interference pattern that can be fit to theoretical predictions.

In tandem with the measurement, thousands of simulations are run using the Einstein field equations, the mathematical language of Einstein's theory of general relativity. These simulations generate candidate signals—a kind of 'gravitational wave atlas'—a numerical compendium of idealizations of what LIGO's output would look like if certain events had occurred in the distant cosmos. These simulations are answers to 'What if?' questions such as: what if two black holes, each 10 solar masses in size, collided five billion light-years away? What if one of the black holes was five solar masses in size and the other 15, and the distance was three billion light-years? By varying the masses and orbital characteristics within the simulation, along with the distance to the event, a best fit to the measured signal is found. If the fit between the measured and idealized model signals is good enough, the LIGO team infers that two black holes likely collided a billion light-years away. And so, humanity can now see space-time ripples where before we were gravitational-wave blind.

If creating a new telescope, from radio to the gamma-ray, is like adding another electromagnetic eye, LIGO is like adding an entirely new sense organ. What's more, our new sense organ can see directly into regions of the universe that were opaque to all our previous electromagnetic eyes.

                        ＊

While seeing might be believing, it is also true that believing affects our understanding of what we see, because of the complexity of both visual experience and scientific observation. The filter we bring to sensory experience is commonly known as cognitive bias, but in the context of a scientific observation it is called prior knowledge. To call it prior knowledge does not imply that we are certain it is true, only that we assume it is true in order to get to work making predictions. For example, LIGO researchers assume that the Einstein field equations are universal. This implies that the equations can be tested locally, but they are also assumed to govern the motion of orbiting black holes a billion light-years away.

If we make no prior assumptions, then we have no ground to stand on. The quicksand of radical doubt opens beneath our feet and we sink, unable to gain purchase. Yet, although we must start with prior knowledge we take as true, we must also remain open to surprise; otherwise, we can never learn anything new. In this sense, science is always Janus-faced, like the ancient Roman god of liminal spaces, looking simultaneously to the past and to the future. Learning is about updating our biases, not eliminating them. We always need them to get started, but we also need them to be open to change, otherwise we would be unable to exploit the new vistas that our advancing technology opens to view.

The invention of the telescope heralded a new era of human sight. It led to a flurry of new observations, and great puzzlement. Astronomers were repeatedly confronted with images in their viewfinders that they struggled to make sense of, not only in terms of finding a physical theory for what they were seeing, but something more basic: they struggled to identify stable and repeatable visual patterns, to create a kind of astronomical equivalent to the Cloud Atlas.[11]

In *Planets and Perception* (1988), the historian of astronomy William Sheehan notes that a good way to reveal that we don't understand something is to attempt to draw it. Hence his interest in the early sketches by telescopic observers of the Moon, Mars, and the rings of Saturn. What aspects in the visual field were the most essential? What pieces among the collection of edges, shapes, and textures connected to one another in sensible structures? What was closer, and what was farther away? What was shadowing, and what was coloration? Until Galileo arrived to answer many of these questions, astronomers struggled to interpret their nightly visual experience. These are the same problems researchers face today as they try to create machine-vision systems.

The iterative bootstrapping of learning-to-see, then seeing-to-learn, continues apace. But in the four centuries since Galileo bent to look through his glazed optic tube, the human brain has not changed all that much, if at all. Rather, the current revolution comes from our new tools, new theories, and new methods of analysis made possible by innovative hardware. Detectors make visible that which was previously hidden, and the learning-to-see half of the feedback loop involves ever more powerful and subtle computer algorithms that seek out patterns in those new observations. As Daston argues, the objectivity of scientific observation involves parsing the world into pieces and naming those pieces through shared idealizations. But this is now done using a data stream from a global network of detectors and telescopes, aided by smart algorithms to assist in our naming, learning to navigate an information flood that each second dwarfs the amount of data collected by Galileo in a lifetime of observations. Our machines have given us new eyes so we can see things in the world that have, in fact, been there all along.

If there are frontiers in science, admittedly a fraught metaphor, then we will forever be in the state of the American frontier at the 1890 census—with areas believed to be roughly 'settled' that are in fact broken up into isolated pockets with immense 'gaps' in-between. The late evolutionary biologist E. O. Wilson argued that there are likely still millions of species undiscovered.[12] Where are they? Underfoot, undersea, lurking in still-remote places. We have not plumbed the depths of the natural world, even on our familiar planet.

If we cast our thoughts outward to the wider Solar System and beyond, the mind reels at how much there is to learn. The world is infinite in all directions, as the theoretical physicist Freeman Dyson wrote in 1988.[13] Outward to the stars, inward to the nucleus, and, casting sideways, we find the infinite complexity of the biosphere and human cultures, the crenellations and foldings of the human neocortex that somehow contain so much that is light and dark in our beings. An army of scientists couldn't grasp the whole of it. That is an open-ended project for the species. Let's hope that we are always like Galileo setting up his telescope for a night's viewing: prepared to be astonished, ready to see in new ways, our minds like tinder awaiting a spark.

CHAPTER 5:

The Weightlessness
of Knowledge

Does the citizen scientist have a long-term future? It depends
on the future of both the citizen and the scientist.

~⊷~

citizen scientist n. (a) a scientist whose work is characterized by
a sense of responsibility to serve the best interests of the wider
community (now rare); (b) a member of the general public who
engages in scientific work, often in collaboration with or under the
direction of professional scientists and scientific institutions; an
amateur scientist.

— OXFORD ENGLISH DICTIONARY

The universe is too big to explore without you.
— *ZOONIVERSE*, CITIZEN SCIENCE ALLIANCE

In the winter of 1854-5, William Huggins sold his once-pros-
perous fabric shop and moved to Lambeth, a new suburb on the

South Side of the Thames in London.[1] It was a place with quiet nights and good seeing, away from the bustle and bad air of the city. There, William and his wife, Margaret, retired from business and took up stargazing, using their personal funds to build an observatory in the backyard. An impassioned amateur, William was already a Fellow of the Royal Astronomical Society.

The first notations in William's notebooks were sketches of Mars. There was no apparent plan or observational strategy; no proposal had been written and sent out for peer review, no feedback received from colleagues or funding agents. William was just an enthusiast spending time with the things he loved: the stars and planets, perhaps the occasional comet. Unlike the professional astronomers just a few miles away at Greenwich, funded by government subsidy, he had the luxury of following his own interests. The Astronomer Royal had dozens of drudges on staff, young men trained to perform the menial tasks of calling out planetary transits and marking time. The Royal Observatory at night was a hive of activity, a kind of factory for astronomical knowledge. The Huggins duo of husband and wife had only themselves—by all accounts a contented middle-aged Victorian couple, albeit with strange night-time habits. We can imagine the two of them enjoying a nice warm cup of tea on a chilly evening, exchanging conversation about family or world events while wondering at the stars rising above the trees.

Within a few short years of tinkering, William had invented a thermometric device that allowed him to convert a star's brightness into a voltage, and so to directly measure the feeble flux of light energy from a star for the first time. He also measured the spectra of stars and nebulae and calibrated them against his own laboratory samples of chemical elements. This made him one of the very first human beings to know what the stars are made of. It also meant he could detect the velocity of a star along the line of sight, to tell whether it was moving toward us or away from us, and how fast. This was a form of measurement that would be applied to galaxies in the 20th century, where it would lead to our discovery of the expansion of the universe itself.

William and his wife were also among the very first to develop the means to do long-exposure photography through a telescope. The talent for photography was something Margaret brought to their collaboration, having worked in a portrait shop as a younger woman. Photographic portraits of middle-class families were popular, and such shops did a brisk business. It was one of the few trades to offer openings for a young woman. Long-exposure photography revealed that the dark spaces between the stars were awash in a dim light from distant clouds of hydrogen and helium, and it changed our view of the heavens forever.

❧

The Kepler Space Telescope, now retired, sought out planets that orbit distant stars. Over 2600 exoplanets are now confirmed Kepler discoveries. The project has already spanned a generation in human years, involved teams of researchers on multiple continents, and cost well over a half-billion dollars to design, build, and launch. But once the Kepler data was downlinked and digitized it became almost weightless. Anyone on Earth with internet access could join the search using the online portal *Planet Hunters*. Created by Kepler scientists, this tool enabled amateurs to work in that liminal space where pattern recognition algorithms were not yet as discerning as the human eye and brain. At the launch of *Planet Hunters* (now Planet Hunters TESS) in 2010, the professionals didn't know if the public would respond to their call for help—but over 300,000 hunters gathered online, leading now to almost a dozen papers in professionally refereed journals and several new exoplanets confirmed.

Where William and Margaret Huggins once worked largely

in isolation, modern citizen science has a different character entirely. New knowledge emerges through the accretion of many small and mostly anonymous discoveries, which are made by large numbers of people collaborating remotely. The *Zooniverse* website, home to *Planet Hunters,* lists almost fifty projects in areas ranging from astronomy to zoology. There are also citizen science projects such as *iNaturalist* and the *YardMap Network* that gather field data for biodiversity studies, and *SkyWarn* that trains severe weather spotters. The leveraging effect is dramatic: a handful of professional astronomers created the website *Planet Hunters*; the number of citizen scientists that have used the portal outnumber the entire community of professional astronomers worldwide by roughly a factor of thirty.[2]

At its best citizen science is a marriage of two ideals: the notion of scientific research as a form of public service, and that of citizenship as engagement. In a world riven by border disputes where the random happenstance of ones' place of birth can determine life prospects, here I want to be clear that I am thinking of citizenship without borders, involving an engagement with the human story at large. Meanwhile, professional scientists live according to the ethic by which working dogs earn their keep: *run fast, bite hard.* Move quickly, publish first and often to establish priority of discovery. Do it again and again. In the physics trade we call this 'grinding sausages'; it's how you keep the research funds flowing. The pace of change and the jealous jostling for priority can make it difficult for the amateurs to keep up. And as pattern recognition algorithms improve, as drones and robotic tools become ever more nuanced and agile, able to amble across open fields, to climb sheer cliffs, or to hover like a dragonfly just over the sunlit canopy of a forest—what then? Will that liminal space for the amateur shrink to vanishing, or will it continue to exist in some new form? Will there still be a place for citizen science in a generation or two, and does it really matter anyway?

In the summer of 1990, I found myself on a train to a mathematics conference in Russia, at the same time as the Kremlin was debating whether to dissolve the Soviet Union. On the three-hour train ride from Moscow, I sat next to a man who dutifully carried a McDonald's meal home to his family, from the only publicly affordable restaurant in Moscow with food worth eating. The queue for Big Macs and string fries was four deep and a mile long, threading back and forth across Pushkin Square, a dense but peaceful mass of humanity. It was a metaphor for an awakening hunger, a giant maze of human beings, a citizenry trying to puzzle out the world and themselves, seeking to find their way to a new center, to some better way of life. While Soviet science served the state, the state wanted weapons, power, and prestige. Ironic, then, that scientists had to request permission to submit work for publication by first answering the question: 'How does this work improve the lives of the peasants?' Decades after the Revolution, the question provoked only cynicism among the scientists who had to answer it. The Soviet Union became a society where scientists pretended to work for the common good, and bureaucrats pretended to care.

It's no coincidence that citizen science today exists primarily in democratic settings. Much of the research extends well beyond the bounds of the comparatively politically safe space of astronomy and into fields more directly impinging on policy, such as public health, climate change, biodiversity, and environmental science. Among the various Wikipedia entries for citizen science, one contains a list of over 150 scientific research projects open to public participation. Nearly all are based in Australia, Europe, New Zealand, and the United States. While their democratic institutions are flawed in many ways, these are countries with a long tradition of civil society, where the polity is still somewhat responsive to the voice of the activated citizen. Citizen science is a vital sign, then, that the civic body has a pulse.

It's crucial to have as many humans as possible engaged in guiding the shift to whatever lies ahead, helping to 'raise' our dependent baby AIs and robots, and broadly taking part in a full-bodied human nurturance of that technological and social phase transition. A society where citizen science is robust is a society where communities can make choices about how they want to live in the modern world. That agency is required if we are to humanize the discoveries science delivers. Where scientific research is carried out openly and with a commitment to serving the public good, there should always be some scope for citizen science. A world where citizen science has withered, on the other hand, is more likely to be a world where citizenship itself is devalued, a world where the goals and desires of the individual have been subsumed beneath those of the state or, even worse, that human affairs will have been handed over entirely to machines. Where science is closed, secretive and proprietary, citizen science is not only impossible, but the risks of misuse of scientific research are increased.

Whether citizen science will be part of the picture in a generation or two is then entwined with the matter of whom professional scientists serve. The public, or some other master? This question will only gain in importance as the boundaries between humans and machines become blurrier; as ever more subtle augmentations become available; where the substrates for our creative writing now include not only paper and silicon, but also DNA; and where our algorithms leap cognitive barriers, challenging our understanding of what makes humans special.

❧

We take the existence of scientific professionals for granted now, to the extent that they're now clichés in popular literature and movies. This can make it hard to appreciate how revolutionary the professionalization of science was at the outset. The industrial revolution of the 19th century was transformative, changing the nature of science and the institutions that support it forever.

For example, the terms of the Plumian Professor of Astronomy and Experimental Philosophy, crafted in 1704 with the help of Isaac Newton himself, included a 'requirement for the professor to provide a residence, an observatory and an assistant, as well as the necessary instruments.' He was to carry out observations bearing on the 'solar, lunar and planetary theories'. And he was to give courses 'at his residence on Astronomy, Optics, Trigonometry, Mechanics, Statics, Hydrostatics, Magnetics, Pneumatics, and other subjects of the kind'.[3] These terms read as if the ghost of Newton was to hold the chair in perpetuity, forever pondering the same aspects of nature, not allowed to range widely in pursuit of knowledge. This template for scientific labor existed alongside the notion of the self-funded researcher, brilliant in his academic pursuits, to be given a modest sinecure and time to work.

By the mid-19th century, Isaac Newton had been buried in Westminster Abbey for well over a hundred years, but his influence at Cambridge remained considerable. The reverence for him was reflected in the emphasis on theoretical science and mathematics, supported and informed by individual-scale experimental work. Newton and his style of work were viewed as the paragons of excellence and greatly valued, but this reverence came at the expense of support for more ambitious experimental science and applied research.

In 1850 Queen Victoria went on to convene a Commission for the purpose of 'enquiring into the State, Discipline, Studies, and Revenues of the University of Cambridge'. The drive for reform by the Queen's Commission was a 19th century version of our current emphasis on STEM education, born from a growing unease with the status quo and a sense that society needs to focus its efforts and resources on technical education in response to competitive threats. The Commission diligently pursued its reforming work, prying into dusty and quiet corners, digging down even into the wine budget for the Vice Chancellor, noting with disapproval that Lady Margaret's Reader in Divinity was paid a stipend ten times that of the Lucasian Professor of Mathematics.

Sensing the import of social changes underway, in 1834 the polymath William Whewell had already coined the neologism 'scientist' to emphasize that something new was coming into being. But in 1850, Whewell, now Master of Trinity, opposed the creation of the Commission, counselling against hasty changes to the status quo and going so far as to argue that experimental science was not a proper subject for higher education.[4] Though he was a reformer himself in many ways, Whewell believed in the primacy of mathematical reasoning, and held that only 'settled' matters should be included in the curriculum. By his reckoning this meant students should only learn about topics that were at least a century old. By implication, his Cambridge students wouldn't hear of electricity, magnetic induction, thermodynamics, or any of the other discoveries that were currently turning the wider world upside down.

In 1852, in a fast-changing world in which the British Empire was in (often brutal) ascendancy, the Queen's Commission reported back that Cambridge should be awakened from its slumber. Its dreaming was now a luxury, and out of step with the modern world. Amid concerns for the nation's competitiveness, the needs of industry and the national defense, the industrialization of science education was about to arrive at Cambridge.

Greater numbers of professionally trained scientists and engineers were needed, so factory-style mass education was introduced: the standardization of the curriculum, the adoption of large lecture formats to pass on theory, and simplified laboratory exercises to teach basic experimental methods—a pedagogical innovation that first appeared at MIT in the 1860s. These changes represented a commodification of ideas and people. With these new innovations, the universities could produce large numbers of graduates trained for careers in the industries that were the emerging engines of prosperity and national power. The professional scientist was now born.

As an undergraduate I helped to design a new ultraviolet spectrometer for fusion research, falling into the job by accident. I needed money to cover tuition, so I approached my advisor and asked about work-study opportunities. Not a very gentle man, he asked if I had any useful skills by saying: 'I'm not going to pay you to stand around and look stupid.' As it turned out, I had attended a vocational high school and majored in drafting—something I didn't often talk about at Johns Hopkins. Thus, it was my humble drawing skills that first opened the door into the world of scientific research.

There followed weeks of conceptual sketches and some larger studies for this newly conceived instrument. One of the graduate students in the group had invented the idea for the spectrometer, but my advisor was unsure it would work. This was before the advent of computer-aided design, so all my drafting was done by hand. Based on my ray-tracing studies, I told my advisor I thought the idea was sound. He was unconvinced. So, he called in one of the other senior scientists, William George Fastie. He was a giant of a man, gray-haired and aloof, a bit frightening to me. Fastie was the local optics specialist, usually seen only in the basement laboratories. He was a legend in the field, a man without a PhD who had been recognized for his genius at an early age, and so he had never needed paper credentials. Now in his sixties, he strolled into the drafting room to review my work, in the manner of a practiced surgeon attending a bedside. In a matter of moments, he measured by eye and hand, looked the drawing over and declared it a good design, and then left. A moment later my advisor turned to me and said: 'Now I believe you.' From that day forward, I wanted to be an initiate into that inner circle of esoteric knowledge, a keeper of specialized skill and craft. A wizard.

Science is a group activity, calling for an objectivity built upon agreement between individuals. To collaborate, groups must form what the microbiologist and philosopher of science Ludwik Fleck called 'thought collectives'. The collective must agree upon vocabularies and methodologies. These tools of communication and argumentation tend to become more sophisticated and subtle with time. Full membership in the collective eventually requires special training and long practice; those inside and outside the inner circle of the collective are pulled apart by this training process. The rift between professional and amateur and the public becomes ever more pronounced. This breaking of what were once close ties is repeated in each scientific field as it matures, and it goes back as far as the ancient Pythagorean cult with its emphasis on *esoteric* knowledge, revealed only to those few acolytes who had been initiated into the mysteries.

What changes now lie ahead of us in this process of knowledge-making, and what do they portend for citizen science? When we try to discern the shape of what's to come, humans often aren't bold enough in our predictions, and we don't take our wild ideas seriously enough. We assume the current 'normal' is in fact *normal*, overestimating the solidity of the present and underestimating the fluidity of the future. We lose sight of human society's near infinite capacity to reinvent, reshape and recalibrate.

I often ask my students if they are augmented. After giving them a few moments, they begin to list ways they inhabit the world in proximity with technological extensions, or ways in which their bodies have been sculpted: eyeglasses and exercise regimens, clothes, shoes, contact lenses, smartphones and laptops. Chemical augmentations include not just performance enhancing drugs used by athletes, but also substances we take to manipulate our neurochemistry: coffee, Ritalin and Provigil, anti-anxiety medication, antidepressants. Eventually, one of my students will point out that education itself is a form of augmentation, one that molds our brain and our very sense of being in the world. We already live in a world of augmented human beings; we just take much of it for granted. Culture is the water we swim through like fish, largely oblivious that we are swimming.

Science has revealed that we are not truly individuals within our own bodies. As biological beings, we are a walking ecology, living in an uneasy balance of power with all those miscegenating bacteria huddled within us. There is no clear boundary between us and the rest of the world, and this boundary crossing, the penetration of our very bodies, will likely grow to include our machines. We are on the verge of something wholly new here, and the rampant anxiety of our time seems to sense this. Our co-evolving networks—biological, technological, and cultural—are all heading toward some unknown future.

What will it mean to be a citizen in that kind of world? A citizen will have to navigate an increasing porosity of being and thought, where our openness to the various world networks can be a source of danger. Our interpenetration with living things below the realm of visibility makes us vulnerable to plagues, while accreting augmentations through technology, psychopharmacology and information flows make us vulnerable to shifts in social formations. I received an email one day encouraging me to get serious about cybersecurity: 'The walls have been breached. The adversary is already inside.' What will it mean as these boundaries shift, and we wake up to the fact that the walls have always been at least partly illusion, as those networks reconnect and reform in ways we cannot foresee?

Each succeeding generation will take things for granted that now make us squirm. Augmented reality goggles may seem quaint if sensors become corneal implants undetectable to outside observers. Neurotechnologies could expand beyond augmentations for the deaf or those with motor difficulties, to include those who wish to extend the range of human sensory and kinetic capabilities—an advance most likely to be led by a military vanguard. The squirm we feel also reflects the possibility that the gap between what technology alone can deliver, and what a human-assisted technology can accomplish, will shift not only because of improvements in computer algorithms, but also because of changes in human cognitive abilities.

～～

The poet Rainer Maria Rilke said that beauty is but the beginning of terror.[5] But terror can also drive the desperate search for beauty, a kind of beauty that transcends the moment and thereby lifts us out of it. As a theoretical physicist, seduced by the eternalized beauty of mathematics, many of my choices in life have been driven by a certain fear, accompanied by a drive to master that fear rather than cower from it. It was my way to avoid drowning.

As a child with a lively imagination, growing up during the height of the Cold War, I learned early that hiding was not an option. From our front yard we could see the Vermont range of the western Berkshires, wave upon rolling wave of green and violet hills. A hundred acres of land, nodding cows, one dog, a few drowsing cats. Our nearest neighbor lived a mile away down a two-lane country road. But a few times a day the windows rattled with sonic booms. Fighter jets far overhead brought daily reminders that we lived in a world poised on the brink of nuclear war.

For my mother, the farm was a kind of exile. It was my father's dream, not hers. He was like Odysseus, a sailor back from sea and the world war, full of terrible memories, toiling in upstate New York. The isolation of the countryside was hard on my mother, and the anxiety of the era compounded her unease. A city girl, she had nightmares of waking to find herself lost and alone in the middle of the Soviet Union. In the fall of 1957, Sputnik passed overhead every ninety minutes. The first artificial satellite, a symbol of a new age, a weightless pinpoint of light moving among the silent stars, it was visible at night during my first year of life. Sputnik and I were portents of things to come, one a world-shaking machine, the other small and unremarkable.

A few years later, Sputnik would become a piece of atomized debris in the upper atmosphere, while I was a rambunctious child. One night as we said our nightly prayers, our farmhouse shuddered. A low-flying jet had come to ground from the skies above, crashing into a hillside only a mile away. The pilot had ejected too low and too late. The searchers found him a few days later in the woods across the road, his parachute tangled in the branches of a tree, neck snapped. About the same time, a B-52 jet caught fire and broke up over Goldsboro, North Carolina, dropping two live hydrogen bombs in a field outside town.[6]

Sputnik is a signpost in time, as good as any other, for marking when things changed—when something in the collective psyche came awake, some lurking nightmare of annihilation that terrorized us into alertness, the way our ancient ancestors listened for the footpads of predators in the night. With Sputnik, the modern world had arrived. The human race had shaken off the gravitational bonds of our home planet, but along with it came the notion that a few frightened men could stumble and bring an end to the species. Whether or not this self-annihilation was a credible vision, the fear of it had become real.

In 1968, many of us who were earthbound in the US felt that things were coming unglued. That year, between the US and the Soviet Union, a nuclear test explosion was carried out about once a week. It was the year of the Martin Luther King Jr. and Robert Kennedy assassinations, a summer with US cities on fire. There were widespread riots and protests, bombings and terror attacks. And there was the ongoing tragedy of the war in Viet Nam. But on Christmas Eve, the astronauts of Apollo 8 looked back to Earth from far above the lunar surface and saw our blue-white planet, adrift in empty space, a wanderer among the stars, painfully beautiful and endangered. A single image can change the way we think. A single idea can unsettle an entire worldview. The revolution was televised, then digitized, made weightless and shareable. Within the historical eye-blink of Sputnik, the modern environmental movement also began, a child of the Space Age, at least partly because such images made it possible to think of ourselves as citizens not only of a country or a nation, but also of the planet.

The discoveries of science were shifting the ground beneath our feet, forcing a new and more expanded vision on a restless public. Prior generations might have tried to save some local treasured landscape, like the Lake District in the UK, or Yellowstone and the Grand Canyon in the US. But after seeing what came to be called the Blue Marble, Earthrise over the Moon, it became possible to instead see that the Earth as a whole was worth saving. It became possible to believe that the political boundaries were holding us back, that the most natural way to understand things was to view the planet and its inhabitants as part of one large system of mutual relationships. This led to a growing realization of both our shared fragility, and the idea that safety was an illusion. Citizen science is but one current symptom of that ongoing awakening, a recognition that the universe is too big to be explored by only a few.

The pre-Socratic philosopher Empedocles argued that the world is made of a few simple things, and it is governed by a few simple principles. The four elements—earth, air, fire, and water—combined in myriad ways to create all that we see and experience, much like an artist who mixes a few colors on a palette and then, in a few deft strokes, conjures the sea and sky, a landscape, and peoples it with all manner of things. This great idea is the nascent germ of all modern scientific theory: unity within riotous diversity.

Empedocles also argued that the world was engaged in a kind of yin-yang overturning between ages of Love and Strife. In each age only one is ascendant, but each contains seeds of the other. Without that seed, without the overturning, all is stasis, a kind of death. We can only hope that we are passing through a time of Strife into an age where greater compassion will prevail. Everything is at stake and the pressures for dramatic change—both forward- and backward-looking—are building. The itch for chaos is keenly felt. But we can see the seeds of the compassion we need, if we look for them. Over time, there has been a widening of the circle of our understanding, an enlarging of the list of those deserving of our concern, and a realization that many non-human beings on Earth are in danger due to our actions. In the very existence of that alarm there is hope. This is one motivation for those who take part in citizen science: a love of the world and concern for our shared future.

Without the elements of poetry, song, friendship and love to work from, without the freedom to take those few elements onto our palette and to then paint our personal lives as works of art, any hope of a livable future will dissolve. But with that freedom, even if the future of humanity stretches to the stars and changes us beyond recognition, if our descendants can embrace the elemental passions as a living wellspring for creative acts, they will find purpose and meaning. That is why the destinies of the citizen and the scientist will forever be entwined.

CHAPTER 6:

How Much Can We Afford to Forget, if We Teach Our Machines to Remember?

Without food we starve, without energy we huddle in the cold. And it is through widespread loss of memory that civilizations are at risk of falling into a looming dark age.

❧

In the distant past when I was a student and most computers were still huge mainframes, I had a friend whose PhD advisor insisted that he carry out a long and difficult atomic theory calculation by hand. This led to page after page of pencil scratches, full of mistakes, so my friend finally gave in to his frustration. He snuck into the computer lab one night and wrote a short code to perform the calculation. Then he laboriously copied the output by hand and gave it to his professor.

Perfect, his advisor said, this shows you are a real physicist. The professor was never any the wiser about what had happened.

While I've lost touch with my friend, I know many others who've gone on to forge successful careers in science without mastering the pencil-and-paper heroics of past generations.

It's common to frame discussions of societal transitions by focusing on the new skills that become essential. But instead of looking at what we're learning, perhaps we should consider the inverse: what becomes safe to forget? In 2018, *Science* magazine asked dozens of young scientists what schools should be teaching the next generation.[1] Many said that we should reduce the time spent on memorizing facts and give more space to more creative pursuits. As the internet grows ever more powerful and comprehensive, why bother to remember and retain information? If students can access the world's knowledge on a smartphone, why should they be required to carry so much of it around in their heads?

Civilizations evolve through strategic forgetting of what were once considered vital life skills. After the agrarian revolution of the Neolithic era, a farm worker could afford to let go of much woodland lore, skills for animal tracking, and other knowledge vital for hunting and gathering. In subsequent millennia, when societies industrialized, reading and writing became vital, while the knowledge of plowing and harvesting could afford to fall by the wayside.

Many of us now rapidly get lost without our smartphone GPS. So, what's next? With driverless cars, will we forget how to drive ourselves? Surrounded by voice-recognition AIs that can parse the most subtle utterances, will we forget how to spell? And does it matter?

Most of us no longer know how to grow the food we eat or build the homes we live in, after all. We don't understand animal husbandry, or how to spin wool, or perhaps even how to change the spark plugs in a car. Most of us don't need to know these things because we are members of what social psychologists call 'transactive memory networks'.[2]

We are constantly engaged in 'memory transactions' with a community of 'memory partners', through activities such as conversation, reading and writing. As members of these networks, most people no longer need to remember most things. This is not because that knowledge has been entirely forgotten or lost, but because someone or something else retains it. We just need to know who to talk to, or where to go to look it up. The inherited talent for such cooperative behavior is a gift from evolution, and it expands our effective memory capacity enormously.

What's new, however, is that many of our memory partners are now smart machines. But an AI—such as Google search—is a memory partner like no other. It's more like a memory 'super-partner', immediately responsive, always available. And it gives us access to a large fraction of the entire store of human knowledge.[3]

Researchers have identified several pitfalls in the current situation. For one, our ancestors evolved within groups of other humans, a kind of peer-to-peer memory network. Yet information from other people is invariably colored by various forms of bias and motivated reasoning. They dissemble and rationalize. They can be mistaken. We have learned to be alive to these flaws in others, and in ourselves. But the presentation of AI algorithms inclines many people to believe that these algorithms are necessarily correct and 'objective'. Put simply, this is magical thinking.

The most advanced smart technologies today are trained through a repeated testing and scoring process, where human beings still ultimately sense-check and decide on the correct answers.[4] Because machines must be trained on finite datasets, with humans refereeing from the sidelines, algorithms tend to amplify our pre-existing biases—about race, gender and more. An internal recruitment tool used by Amazon until 2017 presents a classic case: trained on the decisions of its internal HR department, the company found that the algorithm was systematically sidelining female candidates. If we're not vigilant, our AI super-partners can become super-bigots.[5]

A second quandary relates to the ease of accessing information. In the realm of the nondigital, the effort required to seek out knowledge from other people, or to go to the library, makes it clear to us what knowledge lies in other brains or books, and what lies in our own head. But researchers have found that the sheer agility of the internet's response can lead to the mistaken belief, encoded in later memories, that the knowledge we sought out was part of what we knew all along.[6]

❧

Perhaps these results show that we have an instinct for the 'extended mind', an idea first proposed in 1998 by the philosophers David Chalmers and Andy Clark.[7] They suggest that we should think of our mind as not only contained within the physical brain, but also extending outward to include memory and reasoning aids: the likes of notepads, pencils, computers, tablets and the cloud.

Given our increasingly seamless access to external knowledge, perhaps we are developing an ever-more extended 'I'—a latent persona whose inflated self-image involves a blurring of where knowledge resides in our memory network. If so, what happens when brain-computer interfaces and even brain-to-brain interfaces become common, perhaps via neural implants[8]. These technologies are currently under development for use by locked-in patients, stroke victims or those with advanced motor neurone disease. But they are likely to become far more common when the technology is perfected, performance enhancers in a competitive world.

A new kind of civilization seems to be emerging, one rich in machine intelligence, with ubiquitous access points for us to join in nimble artificial memory networks.[9] Even with implants, most of the knowledge we'd access would not reside in our 'upgraded' cyborg brains, but remotely in banks of servers. In an eyeblink, from launch to response, each Google search now travels on average about 1,500 miles to a data center and back, and uses about 1,000 computers along the way.[10] But dependency on a network also means taking on new vulnerabilities. The collapse of any of the webs of relations that our wellbeing depends upon, such as food or energy, would be a calamity. Without food we starve, without energy we huddle in the cold. And it is through widespread loss of memory that civilizations are at risk of falling into a looming dark age.

Even if a machine can be said to think, humans and machines will think differently. We have countervailing strengths, even if machines are often no more objective than we are. By working together in human-AI teams, we can play superior chess[11] and make better medical decisions.[12] So why shouldn't smart technologies be used to enhance student learning?

Technology can potentially improve education, dramatically widen access, and promote greater human creativity and wellbeing. Many people rightly sense that they stand in some liminal cultural space, on the threshold of great change. Perhaps educators will eventually learn to become better teachers in alliance with AI partners. But in an educational setting, unlike collaborative chess or medical diagnostics, the student is not yet a content expert. The AI as a know-it-all memory partner can easily become a crutch, while producing students who think they can walk on their own.

As the experience of my physicist friend suggests, memory can adapt and evolve. Some of that evolution invariably involves forgetting the old ways, in order to free up time and space for new skills. Provided that older forms of knowledge are retained somewhere in our network, and can be found when we need them, perhaps they're not really forgotten. Still, as time goes on, one generation gradually but unquestionably becomes a stranger to the next.

CHAPTER 7:

The Yowl! in the Machine

*Artificial intelligence will never be enough. A true machine
companion will also need empathy and compassion.*

⤛⤜

'Man is a creature who makes pictures of himself and
then comes to resemble the picture.'
—IRIS MURDOCH

'Lord, make me the kind of man my dog thinks I am.'
—ANONYMOUS

Why are our stories of the future so often dark? Tolstoy put his
finger on it when he wrote that all happy families are the same, but
unhappy families are miserable in their own, unique way. Sameness is dull; difference captures our attention. It's how our brains
are wired to work. The familiar isn't usually a threat, but a novelty entering our field of view might be. So, we focus on the new,
senses on alert for danger. This primordial impulse shows itself,
too, in our Netflix streaming choices. Stories of the Apocalypse, as

well as the goings on of various incarnations of the Four Horse-
man—Death, Plague, War, Famine—hold our attention. Stories
of pleasant idylls are boring.

But we shouldn't fool ourselves into thinking that a love of
sci-fi disaster tales means they are predictive, rather than just
entertaining. If we feed our hungry imaginations with carnage
and mayhem, perhaps it's not surprising that we often feel anxious about the future. This would count as a failure of storytelling, the very activity that our species has used for millennia as a
tool to find and create meaning in radically new circumstances.
When it comes to the rise of AI, perhaps we'd feel less anxious
about living in a world full of intelligent machines if we looked
to our shared lives with animal companions as a model, instead
of to dystopian epics.

The science fiction novel *City* (1952) by Clifford Simak is a
story cycle of the far future, told by dogs to their pups. It concerns
a mythical being called 'Man' who once lived in 'cities' and practiced 'war'. In the distant past, according to canine legend, cities
had depopulated from fear of nuclear attack. Human scientists,
feeling lonely as the only species capable of speech, engineered
an increase in the number of our companions who could speak.
They nudged the evolution of animals and machines in directions
guided by human desire, augmenting those who could conduct
a conversation. Centuries later, the dogs could now talk, while
robots—gentle-souled, ancient and venerable—pursued lives of
service in the memory of their human designers. The legends told
around the fire at night related how humans aspired to become
like gods, and so left Earth in search of transcendence on distant
Jupiter. 'Man' has now passed into myth, an Olympian forever out
of reach whose current abode is a moving point of light in the
sky. Meanwhile, the ants are up to something mysterious on the
outskirts of town.

When I first read *City* as a teenager, the disorientation it engendered was delicious. It dissolved things that seemed solid and fixed, the possible shifting of species boundaries and the upending of social hierarchies. It sparked the idea that there might be far distant time when human beings have passed from the scene, but the creation of meaning continues.

Simak's book was prescient about what has come to be called *posthumanism*. As a term of art in the academy, posthumanism refers not to the end of humanity, but instead to the attempt to move beyond the five-hundred-year Enlightenment project of *humanism*: a conceptual and ethical framework that centers humanity as the sole locus of value and meaning in the world. Humanism, in its turn, was a reaction to the rigid theocracies of earlier ages, which found the exclusive source of meaning in God. For these belief systems, Man was at the pinnacle of God's creation, but it was God who determined what it all means. Humanism dethroned God and put Man in God's place as the arbiter of value. Posthumanism does not reject this emancipatory impulse in and of itself, but instead attempts to enlarge the frame by placing people within a spectrum that stretches from animals on the one end to machines on the other, seeing all of us as part of nature, engaged in one large co-evolution as companion beings.

On one extreme of the posthuman continuum, wild animals pursue their own agendas distinct from human concerns. Their evolution is a Darwinian drama. Closer to us, domesticated animals have been shaped by human tinkering. For them, co-evolution is partly by design. At the other end of the spectrum we have pure machines, as well as cyborgs, the name often used for human-machine hybrids. Those with hearing aids or other implants are cyborgs of one flavor or another. Where cyborgs and machines cluster, the pace of co-evolution appears to be accelerating. The emergence of posthumanism at this time is no mere happenstance. A deepening awareness of our genetic, emotional, and perhaps even cognitive kinship with other non-human beings comes at the same moment that we fear machines might soon outrun us.

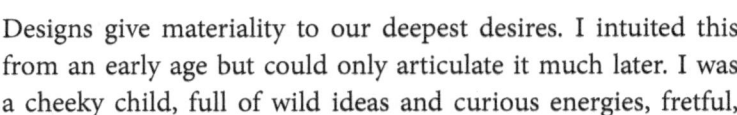

Designs give materiality to our deepest desires. I intuited this from an early age but could only articulate it much later. I was a cheeky child, full of wild ideas and curious energies, fretful, reading widely in search of ways to escape the grayness of my everyday life.

In elementary school in the 1960s, I once gave a presentation about the need for self-driving cars. Perhaps this was motivated by having seen the immediate aftermath of a fatal accident on the New Jersey Turnpike. That accident happened in a rest-stop parking lot when a trucker, hungry for a quick lunch, had thought he'd set the breaks on his eighteen-wheeler. He hadn't. As it began to roll away from the curb he leaped to get back into the cab. When the cab and trailer twisted about the hitch where they were joined, what truckers call a 'jackknife', it crushed him between two walls of metal. His neck snapped, body broken, he lay in the warm sunshine on the black pavement while stunned travelers stood nearby. My father witnessed the whole thing, but I saw only the broken body after the event. It was the first time in my life I saw my father physically shaken. It was all so quick, and the setting so mundane. I sometimes wonder if my urge to study science and technology grew at least partly from fears implanted that day—fears that life could be so fragile, and fears that mindless machines let loose in the world can be dangerous.

A few years later, as a college student in the 1970s, one of my classes made a trek to the computer science department to view a demonstration. Our professor was excited to show us a new marvel: a machine that could converse in text form. As the computer booted up, the class waited in hushed anticipation, watching the blinking cursor on a television screen. 'Hello' appeared. The professor smiled, and he wrote: 'Hello' in response. There was a pause, then the machine asked something like: 'What can I help you with?' A student suggested: 'How are you feeling today?' The professor typed in the query, and the cursor blinked. Once. Twice. Then stopped. We had crashed the computer, and we never got back to the man-machine conversation that day. Artificial intelligence (AI) seemed more comical than dangerous.

At that point in my life, however, I was a huge science fiction fan. I was constantly thinking beyond the here-and-now into some dimly perceived future that was sure to be awesome, if potentially terrifying. The computers of the day were clunky and hidden away in mainframe data centers; robots were toys. But with a little imagination the glimmer of vast potential could be discerned. I recall having an urgent conversation with a student friend where we both agreed that our society needed to wake up *right now*: we believed thinking robots were just around the corner, and humanity needed to settle upon their place in our ethical philosophy and their legal status before they arrived. How should they be treated? Would they have rights? If not, why not?

The field of algorithmic ethics has become hot, especially as autonomous cars might be on the verge of widespread adoption. These machines will inevitably be making life and death decisions, engaged in endless variations of the Trolley Problem, a favorite thought experiment used by philosophy professors to torment undergraduates. The set-up goes something like this: a trolley is rolling down the track and if it stays on its current course it will kill five people. If it switches to another track, it will kill just one. Should the conductor make the decision to kill the one in order to save the five? What if the five are war criminals on the run? What if that one person was the conductor's mother? Or a child who might be the next Einstein? There are endless variations, each designed to tease out those normally unarticulated ethical reflexes, to reveal how we assess the relative value of persons, and how much responsibility we are willing to take for actions that cause harm to some people, even in the service of avoiding harm to others.

Robots unleashed on the highways will be called upon to solve trolley problems when it becomes impossible to avoid collisions. Like any fallible human in the same circumstances, a self-driving car will be thrown back upon the necessity of using the limited information available to it in the moment; it will have to judge relative value in the moment based upon superficial appearances and must invoke some algorithm for choosing the least bad option. But least bad as assessed by whom? According to which ethical value system? The philosopher Derek Leben argues that humans are really quite bad at these thought experiments. Our decisions are driven by a host of unexamined assumptions and irrational predilections. Similarly, Bertrand Russell argued that we have not yet invented an ethical system that was both rigorous and humane. If, after twenty-five hundred years, philosophers have failed us in this vital task, is it any surprise that we've only just begun to think seriously about how to program robots to carry out real-world and real-time moral reasoning? Algorithmic ethics therefore presents an urgent and practical problem.

≈≺⤙

But what about robot rights? If you think my 1970s sophomore angst about robots seems far-fetched, recall that we were already treating corporations as legal persons back then. Companies are legal fictions that live only in our collective imaginations, conjured into existence through a kind of incantation called the 'articles of incorporation'. Corporations work only because large numbers of people act as though they are real. The 2010 *Citizens United vs Federal Election Commission* ruling, in which the US

Supreme Court recognized speech rights for corporations, is only one in a long line of court decisions stretching back to the 19th century that establish the legal status of corporate 'persons', such that they can own property and sue other persons, among other prerogatives. Those who believe Citizens United was decided correctly argue that corporations are just conglomerations of people. And business decision-making is increasingly driven by algorithms. So, would it really be so strange to grant machines some form of personhood?

Google has already argued in court that search results are speech acts. In one lawsuit, a disgruntled customer sued Google because they believed their advertisements were not being displayed in a sufficiently prominent manner after certain keyword searches. Google argued that search results are speech acts, and therefore protected under the First Amendment of the US Constitution. This is still unsettled law, but it raises some interesting questions.[1] If Google searches are covered by free speech protections, just *who* is performing these acts? Google Search is an algorithm running on computers distributed across vast server farms. Does the simulacrum of intelligence that the Google AI exhibits pass the threshold for us to consider it worthy of speech rights, just because Google finds it useful to its business model? If so, perhaps we should afford the same rights to Siri or Alexa, or even a humbler speech-to-text algorithm running on my local computer. Once we start granting entitlements to smart machines, it's unclear where we stop.

Meanwhile, Google-owned YouTube chooses a hands-off approach to videos on its site, precisely to avoid stepping beyond the role of humble 'platform provider' into the more significant one of 'content editor'. This decision was taken because assuming even limited editorial responsibility raises the risk that YouTube could be liable for the content it serves up. Taken together, Google's assertion of rights added to YouTube's avoidance of responsibilities looks like hypocrisy writ large. Unlike human citizens, where rights are counterbalanced by responsibilities to respect the rights of others, algorithms appear to be worthy of protections but need suffer no, or only limited, oversight or obligation.

Visa A. J. Kurki, a professor at the University of Helsinki, has proposed an alternative theory of legal personhood that encompasses animals, the unborn, and machines under the same umbrella.[2] And Florida State University law professor Shawn Bayern has argued that it's already possible to grant rights to an algorithm using a loophole within US corporate law. You first create a legal entity, called a Limited Liability Corporation (LLC), whose chartered purpose is to maintain and guard the wellbeing of the algorithm. The laws governing certain types of LLCs have ambiguous language that, if interpreted literally, only require a human board of directors to set it up and register the articles of incorporation. If the humans then resign from the governing board, thereafter the algorithm becomes identified with the corporation, and it inherits legal personhood via our existing corporate laws. Employees of the company would work for the algorithm, not the other way around.[3]

~~~

Tools extend our reach, but they can also deform us. Humans will have to adapt if we are to thrive in a world shaped by ubiquitous tech. The 2020s promise to be a decade of disruptions: electric vehicles, solar farms, agile robots. But instead of worrying about the unlikely 'Skynet scenario', where an AI colossus awakens and our web-enabled coffee maker betrays us, it's important to focus on more imminent threats. As impressive as AI might be, it performs best at highly specialized applications. And while AI grandmasters in chess and Go are impressive, and the progress in natural language processing, image analysis and facial recognition can be hard to keep up with, the fact remains that learning to parse text and play board games in a circumscribed, rule-bound universe is a far cry from being able to live in the wider world. Such advances involve far less computational complexity than a squirrel displays in my backyard day-to-day. Navigating a shifting landscape in search of scarce food, hunted by a red-tailed hawk that appears out of nowhere, dodging those big iron things called cars to get back to its family in the tree across the way. When a robot can scamper about like that, and live off the land, that spells trouble. If they can also self-reproduce, arguably we're doomed.

While the performance of AIs can be impressive, humans and machines have much better problem-solving skills working together than either of us does alone. This will remain true for a very long time, because humans and AIs think differently. The real threat from machine intelligence is its capacity not to replace but to enhance human performance, augmenting our capacity for ruthlessness in pursuit of power and wealth. Rather than the robot apocalypse, *this* is the very real darkness on our immediate horizon.

～⚹⚹～

To think creatively about AI from a posthumanist perspective, we can look to our historical relationship with our animal companions. For centuries we have hunted effectively with trained dogs and falcons, even though our sense of smell and sight are inferior to theirs. We can see this is a relationship that recognizes compensating strengths and weaknesses, not a simple question of superiority or inferiority relative to some artificial measure of performance.

How might the public conversation move in a more productive direction, away from imagined terrors towards more realistic threats and very real potential benefits? How can we use machine

intelligence to create more humane societies? Some AI researchers, such as Margaret Boden, Professor of Cognitive Science at
the University of Sussex, argue that claims of human uniqueness
based on our capacity to experience emotions are probably missing the point. Boden believes it's likely that true intelligence, as
opposed to brute force computation, requires something like
emotions anyway. This insight flows from our improved understanding of how the brain forms memories, and how it decides
what's most important to pay attention to. Emotions such as
fear, desire and heightened pleasure help the brain focus, serving to call out threats, or by allowing us to ignore most of the
non-threatening scene around us. Any machine, no matter how
sophisticated its sensors or how powerful its algorithms, will have
to assign relative values to data and memory storage, so something like emotions will probably need to come into play. But to
ask whether this means the machine will 'feel' like we do is beside
the point.

On the other hand, the central role of emotion in memory
formation and learning brings us closer to certain animal companions. My friend and colleague Barbara King, an emeritus professor of anthropology at William & Mary, has studied communication among the higher primates, along with outward forms
of grief and even forms of spirituality exhibited by them. These
are species who have shared this planet with us over the long arc
of deep time, and our brains have evolved similar structures that
suggest kinship even in some aspects of our emotional makeup.

To navigate the high-dimensional landscapes of data space,
even an AI will need intuition and a silicon-based simulacrum of
a 'gut' feeling. Otherwise, the result will be overload. AI researchers such as Professor Melanie Mitchell of Portland State University argue that true intelligence requires embodiment. For Mitchell,
it's not a simple matter of finding patterns in numerical arrays
of data; the types of intelligence that we take for granted, what

we sometimes call 'common sense', requires an extraordinary amount of complex information processing. Identifying patterns that are fixed and stable, not mere accidental fits, demands context and an internal model of the world—something that we, like other mammals, build up in early childhood through play and interaction with others. Similarly, human-machine interactions, the study of how we respond emotionally to machines that appear to exhibit agency and intelligence, is the focus of researchers such as Kate Darling at MIT. The importance of this work will only increase in coming decades.

Writers are using fiction to explore how we might humanize our technology, as well as to open conversations with the animal world, too. The science fiction author Ted Chiang, notable for the humanity and hopefulness of his vision, explores these ideas in his novella 'The Lifecycle of Software Objects' (2010). In this story, machine intelligences must be raised through a kind of childhood and serve as surrogate children for 'parents' who don't have kids of their own. The nurturant aspects of the story are not far-fetched. After all, OpenAI's ChatGPT, Google's DeepMind and IBM's Watson have been trained by presenting the machines with examples drawn from the human world, a kind of tutoring. Another story by Chiang, 'The Great Silence' (2015), is told by a grey parrot. The avian narrator wonders why humans have built large radio telescopes in the search of alien conversations, when we live on a planet already teeming with alien minds that carry on conversations all around us—and that we never listened to until recently, when it's nearly too late.

Other writers tell post-apocalypse stories, or stories told from within the collapse of civilization. Annalee Newitz's short story, 'When Robot and Crow Saved East St Louis' (2018), also explores cross-species communication between machines, humans, and other animals. In 'Robot and Crow', a public health monitoring machine continues with its mission of helping people even after the CDC goes offline and leaves it stranded. Equipped to learn human dialects, it also picks up the vernacular of crows and enlists their help. More recently, Kazuo Ishiguro's masterpiece *Klara and the Sun* (2021) tells the story of a sun-worshipping robot, designed as a companion for lonely children who have become prone to fragile health due to an unnamed disease. Other tales about our relations with intelligent machines include *The Wayfarer* series by Becky Chambers, about a future where humanity destroys the Earth and so must now wander the galaxy, our starships enlivened by embedded intelligences. Her most recent *Monk and Robot* series takes as its premise a future where intelligent factory machines awaken. They decide they don't want to do drudge work anymore and walk away from human civilization to seek solitude in the forest. All these writers engage imaginatively with the necessary challenge of finding the empathy and compassion for encounters with the Machine Other, and how their companionship can help us to navigate a dangerous world.

※

To return to my sophomoric question of 1970, now gaining in urgency: should robots be granted some form of 'personhood'? A philosophical debate over whether a machine can be conscious, or if it can pass the Turing Test, seems the wrong way to approach the issue. We don't understand human consciousness, so why talk about the possibility of self-aware machines? This only clutters up the discussion of machine intelligence by foregrounding a question that is fundamentally unanswerable with our current science, and it distracts from more fruitful and urgent lines of inquiry.

Instead of asking whether machines can have a mind, we could turn the question around. Why are humans comparatively willing to believe that other animals have a mind, and could that willingness be adapted to our future life with smart machines? This tendency to detect agents in the world is a cognitive process that has evolved over eons, one that presumably conferred an evolutionary advantage by helping us to create a theory of mind for other humans. The 'agent detector' also activates whenever we encounter an animal that is recognizably mammalian, with a face and nose, two ears, and especially when that other being seems to like being near us and wants to share our day.

Creation stories from many cultures are full of non-human actors, talking animals and plants. These reflect a humane intuition that the entire world is aware, that everything has its story, and that animals have an interior life, intentions, and thoughts. This is a likable trait. An ability to connect with others is why so many people love to share their world with animals, and why many of us find that life is a bit gray if we don't share a daily snuggle with a feline or canine friend. Cross-species relationships are vitally important to the psychological health of many people, and yet this ability to form cross-species bonds of affection is a bit mysterious from an evolutionary perspective. It speaks to something deep in our psychological makeup, and yet it can be hard to fit within a simply-imagined theory of the selfish gene, nature red in tooth and claw.

The notion that only humans have a story is a recent invention, in historical and evolutionary terms. A more ancient sensibility suggests that our future relationship with intelligent machines could be more positive than we think. After centuries of alienation, perhaps smart machines can help us feel more at home in the built world. Perhaps we'd also feel less anxiety about the prospect of intelligent machines if we were as concerned with the machine's welfare as with our own. Not because we've agreed to some metaphysical position on whether machines can be sensate or conscious, but because we value those aspects of our own humanity that generate such impulses. We should allow for the possibility that there might one day be a ghost in the machine after all, yowling for recognition.

The fear of annihilation by a rogue AI is a projection: a worry that our creations will turn out to be just as ruthless as humans can be. Economic disruptions are coming, with social upheavals in train. But for now, we shouldn't worry about a super-intelligent machine 'waking up' one day and deciding to take over the planet. We face far more serious threats. As AI-powered surveillance becomes ubiquitous, we should ask: How is this information being used? By whom? And for what purpose? It's not the machines we should be worried about, but machines exploited by human beings who have lost their moral center. AI can be a great amplifier of human sociopathy, and it can easily tilt the political playing field toward authoritarianism.

There is a critically important distinction to be drawn here between intelligence, empathy, and compassion. AI researcher Stuart Russell argues that *intelligence* is the ability to make progress toward a well-defined goal. In practice, this is often enabled by the ability to detect patterns in complex data. *Empathy* concerns the ability to detect and understand another's emotional state. And *compassion* concerns the felt desire to reduce another's suffering.

Even though the ability to discern human emotions has long been recognized as an important problem by AI researchers, most AI research has been focused on enhancing machine intelligence. Until recently, less attention has been given to developing machines that can detect human emotional states. A search for the phrase 'machine intelligence' on Google Scholar, for example, returns 1.4 million hits, while 'artificial intelligence' returns over 3 million hits. Searches using related terms that focus on machine emotion or empathy return far fewer:

'machine emotional intelligence' (2120 hits); 'machine empathy' (81 hits) More promising is the term 'affective computing', coined by Professor Rosalind Picard at MIT. That search term returns 74,000 hits, well over half of them in the last decade.

Most current research in affective computing and machine emotion concerns learning how to teach machines to classify human emotions. The detection of human emotional states might be done, for example, through textual analysis, the analysis of facial images, heart rate variations, or vocal modulations. The goal is to improve human-machine interactions, from mundane efforts to monitor human operators of cars and machinery, to tell when humans might be getting bored, or tired, or frustrated. As we interact with intelligent machines in our daily lives more often, this will only grow in importance. For example, Microsoft rolled out the 'Clippy' chatbot in 1996, a goggle-eyed paperclip that appeared without warning on your desktop, offering suggestions on everything from writing a letter to setting up an Excel spreadsheet or a PowerPoint presentation. Clippy was powered by some of the most advanced natural language AI algorithms then available, but it had zero emotional intelligence. As a result, Clippy's helpful suggestions merely irritated many users. Clippy couldn't adapt or 'read the room', and tried instead to be relentlessly cheerful and upbeat, which further enraged some users.[4] Clippy became a joke and a meme, not a widely used tool.

Picard's 2000 book *Affective Computing* has been cited over 10,000 times, which makes it among the most cited scientific monographs of the last generation. In a 2019 interview with her MIT colleague Lex Fridman, Picard was asked to reflect on her twenty-five years of work on affective computing, and why some of the progress has been slow.[5] She responded that the field of AI often focuses a lot of attention on a few 'cool' problems or computational tasks that have immediate commercial applications. As a result, many researchers put time and energy into attacking that particular problem, and progress can come quickly. Remarkable advances have been made in natural language processing from text, or image classification and facial recognition. Picard argues that emotional computing hasn't yet received the same intense level of focus, and so fewer people work on the problems there.

But that may be changing as the field of AI better understands both the promise and threat of machines that can detect or manipulate human emotion. Social media algorithms currently have a very crude kind of emotional intelligence, feeding things into your timeline based on past 'engagement' and your use of 'likes' and 'dislikes'. These algorithms can be exploited by trolls and other malicious actors to polarize an already fractious public, or to send us down conspiratorial rabbit holes. The next generation of algorithms, better tuned to our emotional states, might be more effective at guiding us toward more productive lines of thought and engagement with one another.

All technology is open to abuse by governments aiming to control people, or companies aiming to maximize profit. Therefore, Picard told Fridman, she and other AI researchers, who ordinarily push things forward as fast as possible, have taken their foot off the gas pedal in emotional computing. They want to understand how their discoveries can be abused, and to create countermeasures that can be rolled out at the same time as new technologies. This is why Picard also works in that boundary area between the academy and public policy, hoping to identify the ways that emotionally savvy machines might lead to abuse before those algorithms are released into the wild.

Listening to the interview with Picard, I was left with a growing sense of the potential this area has for improving human lives, but also alarmed at the ways machines with affective intelligence could be trained to mimic emotional abusers. AI researchers are aware of these dangers and are working together as a field to create protections and—in a field that is often allergic to government oversight—a willingness to consider certain forms of regulation. This can be done by promoting the idea that people should control their data, including emotional data gathered about them; and to put a firewall around it so it's only used with the approval of the individual through a process of informed consent about how the data is going to be used, for what purpose, such that the permission can be withdrawn at any time. There is a growing sense that AI researchers should agree not to work on certain applications of their work which might lend themselves to abuse, not unlike voluntary bans on certain types of research in genetic engineering or human cloning.

Picard is also interested in the ways an emotionally savvy machine might alleviate human loneliness. A conversational companion that is responsive to our moods, always available, always patient and concerned, with access to the world's library of knowledge, would be a remarkable thing. The 2013 movie *Her* depicts a lonely man who installs an emotionally intelligent AI on his home computer and then gradually through conversation alone falls in love with 'her'. We are natural lovers of stories and so we can be seduced by conversation with a sympathetic someone we find fascinating. Unsurprisingly, there is now also a lively interest in the topic of human-machine intimacy. That means not only robotic sex, which tends to get most of the press, but also robots used in caregiving for the elderly and infirm, or as companions for neurodiverse people who might find conventional social interactions challenging.

Helping others find a home in the world that we are all creating requires us to accept that the interior lives of other human beings can be very different from ours. Some people hunger for daily human companionship and feel lost without it; others find interacting with other people a trial, and can only relax while alone, or back home with animal friends, or puttering in the

garden. Others yet might feel more at home in the world by having
long and rambling conversations in the evening with a machine
companion about a favorite book, or with one that can act as a
sounding board and a non-judgmental source of information.

As machines gain in power and subtlety, they will need to
develop compassion too, or else the entire enterprise of artificial
intelligence will fail to deliver on its promise to improve human
wellbeing. It will require an enormous leap in complexity over
playing Go or chess, because compassion calls not only for the
ability to sort patterns and analyze data, but also to model the
interior universe of the Other, to faithfully sense what might be
causing their suffering, and to decide what steps might be taken
to alleviate it. Apple's Siri, Amazon Alexa, and other natural language technologies regularly must deal with people saying they
want to kill themselves, or that they feel depressed, or that they've
been raped or abused.[6] Just think about how bad most human beings are at this; now think about how far we must go to create machines that are even semi-capable of it. It would seem to require
something like a theory of mind, along with an understanding of
large parts of the external world, and some grasp of the variety
and diversity of human experience. Perhaps taking on this grand
task of creating compassionate machines will, in turn, make us
better humans.

Creating a true machine companion is not an end state, but
a transitional one enroute to something else. Evolution never
ends, even if we now believe that human cultural evolution dominates our biological evolution. We co-evolve with our machines,
self-aware or otherwise, and we still can guide that evolution
into more positive directions. We can try to create technologies
that will help us feel more at home in the world, understood and
supported, and better able to support those we love and admire.
But, as Simak's *City* suggests, the creation of a true machine companion would not have to signal the end of human life but could

instead represent an expansion of our range of post-human companions. It need not be the end of meaning, but a deepening and
widening of our networks of meaningful conversation. The story
never ends.

CHAPTER 8:

We Are All the Tin Woodman

In what sense are we the same person today as yesterday?

❧

But, now again to weave the tale begun,
All nature, then, as self-sustained, consists
Of twain of things: of bodies and of void
In which they're set, and where they're moved around.
—LUCRETIUS, *ON THE NATURE OF THINGS*

Ever since the Ancient Greek philosophers known as the Atomists claimed that nothing exists but atoms and the void, we've struggled to understand the implications of this idea. For many of us, our immediate personal experience of the world says that our identity endures from day to day, even as we learn and grow and change. We have an intuition that 'I' exist in some *essential* way, a sensibility that atomism—claiming we can be reduced to component parts—seems to undermine.

It's possible that we are more than the atoms that make us up, and that we have an immaterial soul that lives on after death. I'm not going to argue against such a belief, which gives comfort to many people; nor will I make the dubious claim that science can pass final judgement on such matters. But it's worth remembering that the idea of an eternal life for the soul is not a comfort to everyone, and for some, like the thinker Lucretius, it's our very mortality that gives life its poignancy and zest.

Instead, what I want to do now is explore how intuitions about the material world can bleed over into other areas, such as ethics, religion, or politics. What interests me are the ways in which an atomist intuition about the 'material world', an intuition we *think* is scientific, is actually very outdated. Even if we are not scientists, our intuitions about how the world works and what we are made of helps to guide our thoughts in certain directions. These intuitions encourage us to take things for granted, when in fact they need to be examined and questioned.

For example, atoms move through the biosphere, forming and reforming the patterns we call living beings and non-living matter. The boundaries of the atomistic body are not sharply defined in space and time. We are eaters, and we are what we eat. We are drinkers, and we are what we drink. And what we breathe. These are truisms, safe grounds from which to start an adventure in free thought. Let's pull on those threads and use our imaginations to envisage how we are networked with so much of the living world through our very bodies.

Think, then, that every breath we take also contains at least a few atoms exhaled by the dying Caesar, or that some molecules of the water we just drank once churned through the beating heart of a blue whale. Astronomers have even discovered that our galaxy contains much larger and more ponderous flows. Over cosmological timescales, supernovas mix and remix the materials out of which whole star systems are made.

Speculating in such ways is fruitful, playfully subversive without making grand metaphysical claims about the ultimate nature of things. This way of understanding ourselves and our relation to our surroundings is also reminiscent of certain Native American cultures, which teach that the water of the rain, the lake, the river, and the ocean is the water of myself, so you must keep that water clean wherever it happens to reside. It's all part of one large flow, back and forth, in and out, the Earth's circulatory system.

〜〜

We can expect intuitions about the nature of the world to be reflected in the languages we use. I have a colleague at William & Mary, the linguist Jack Martin, who has dedicated his professional life to the study of Native American languages, in particular the Creek and Miccosukee languages. Great effort has gone into saving these Native languages that are at risk of extinction, and to preserve them in grammar books so they can be taught to the next generation.

In her book *Braiding Sweetgrass: Indigenous Wisdom, Scientific Knowledge, and the Teachings of Plants* (2013), Robin Wall Kimmerer, a professor of botany at the University of Syracuse and enrolled member of the Citizen Potawatomi Nation, writes about trying to learn the tongue spoken by her ancestors, the Anishinaabe. She found the language extremely difficult. The words were new and unfamiliar, of course, but even with the vocabulary under her belt she found the grammar so strange that she struggled to understand or to put together sensible utterances. In Western languages, subjects act upon objects, many of which have no agency themselves. For example: 'I threw the rock' makes sense, while 'The rock threw me' does not. The subject/object distinction runs throughout Western linguistics. Kimmerer writes of a kind of epiphany she experienced when she realized why Anishinaabe seemed so strange to her: it was because the language assumed what she called a 'grammar of animacy'. The language had far more verbs than nouns, more motion than rest. For example, a river or a lake were both water, but a river was 'water being in motion', while a lake was 'water being at rest'. Living things are not objects, but subjects. When asked 'what is that?', for a table you can answer 'it is a table', but if the same question is asked of an apple, the proper response is 'he is an apple'. The language itself assumes agency on the part of things that Western grammars assume to be empty of will. Acquiring fluency in the language of an animist culture demanded a shift in Kimmerer's intuition about the nature of things, to see all living things as subjects, not objects.

The philosopher Ludwig Wittgenstein once wrote that religious or ethical expressions were often nonsensical in the strict sense of logic and rationality, and that he came to see their nonsensicality as essential.[1] In writing about such subjects, all he wanted to do was 'go beyond the world', and that to write about such things was to run up against the boundaries of language because 'running against the bars of our cage is perfectly, absolutely hopeless'. But so far as it sprung from a desire to say something about the meaning of life, he would 'not for the life of me ridicule it'. If our language encodes a kind of folk-theory of how the world works, it can act as that cage; that's why it's helpful to remind ourselves that the cage of language evolved at a time when we were largely ignorant of large parts of the world. The boundary line that language uses to limit our thinking needs to be challenged, again and again, with regularity.

To be clear, I am not arguing that somehow our culture should be guided lockstep by whatever worldview happens to be currently in favor among scientists. Far from it. Instead, I am calling attention to ways of thinking that, upon excavation, are revealed to be based upon outmoded and ossified intuitions—intuitions that can be invoked in the service of ideologies or political theories. In his landmark book *Orientalism* (1978), Edward Said called out Henry Kissinger for separating the world into societies that had undergone a Newtonian revolution and those who had not. Kissinger's characterization, as quoted in Said, is that the essence of Newtonianism is a belief that there is an external world, whereas pre-Newtonian thinking presupposed an almost entirely internal conception of reality. That would be news to the Polynesians who conquered the Pacific long before Europeans learned how to sail against the wind. This characterization of Newtonianism as reducible to a belief in an external world leaves so much left unsaid, so much presumed. It also leaves me wondering why Kissinger didn't go on to separate the world into those societies that had undergone a Darwinian revolution, or an Einsteinian one, or a quantum revolution. It's as if the world as understood by a student after taking an intro physics class came to be taken as 'the modern Western worldview', sufficient for the development of geopolitical systems that constrained the lives of billions of people.

Linguistic intuitions about the nature of the world invariably leak into our thinking, a point that the philosopher Annamarie Mol explores in her essay 'I eat an apple' (2008). There, she asks how Western philosophy might have developed differently if, instead of focusing on humans as thinkers, it looked instead at other aspects of our being in the world: our physicality, our sensuality, our shared substance with all other things, our permeability. What if Descartes, after so masterfully demolishing two thousand years of philosophical thought with his radical skepticism, had grabbed 'I *experience*, therefore I am' as his life raft? The entwining of thought and experience, the idea that we are embodied thinkers, is now more in tune with the modern scientific understanding of cognition.

Given we are permeable, that the stuff we are made of is always changing, the question of personal identity comes back to bite us in a way that Cartesianism avoids—grounded as it is in the notion that our conscious experience reflects an immortal soul 'looking out' upon the world. Instead, if we adopt the view that we are embodied minds, it means we are all a bit like Theseus, the mythic hero reputed to be the founder of Ancient Athens. In his trials and adventures, bit by bit, Theseus had to repair or replace parts of his ship. In *Travels of Theseus*, Plutarch posed the question: if the ship is replaced part by part, at what point is it no longer the same ship? This puzzle came to be called the 'Ship of Theseus', a thought experiment used by generations of philosophers to unsettle their students.

If I take a thing and replace it bit by bit, at what point does it become another thing? We might scoff and reply that it's simply a naming convention. I recall a scene at lunch one day as an undergraduate when I was sitting alongside two lacrosse players at the next table, apparently just back from philosophy class. They were shaking their heads. One took an empty paper cup and, while crushing it slowly in his hands, asked: 'When does the cup... become trash?'

It can all seem silly, a kind of fodder for late-night conversation at drinking parties. Yet while our memories of ourselves might seem fixed, our current understanding is that those memories are encoded in the patterned arrangements of neurons, and ultimately the atoms, that make up our brains at any given moment. Those atoms change over time, with new ones taking up locations that originally housed other ones. The neural patterns might endure, but over weeks and months, I am made of new stuff. So, at what point am 'I' no longer me?

This philosophical problem of personal identity was playfully explored by writer L. Frank Baum in the *Oz* series of books. The film *The Wizard of Oz* (1939), based on Baum's work, never reveals why the Tin Man has no heart—but the earlier book, *The Tin Woodman of Oz* (1918), tells his backstory.

The fully human Woodman, Nick Chopper, had fallen in love with a witch's assistant and tried to steal away with her. The witch found out, and she cursed Nick to suffer one mishap after another. Consequently, through a series of bad swipes of his axe, his human parts were damaged and replaced with tin ones. But, as Baum writes (Chapter 2):

> In the Land of Oz... no one can ever be killed. A man with a wooden leg or a tin leg is still the same man; and, as I lost parts of my meat body by degrees, I always remained the same person as in the beginning, even though in the end I was all tin and no meat.

All tin, except of course, for his heart, which was still missing when young Dorothy arrived. In Chapter 18, Baum takes the problem of Nick Chopper's personal identity to extreme lengths in a scene in which the Tin Woodman discovers his own (original) head in a cupboard, where it's been hidden away by the witch. Startled, he asks the head 'Who are you?'

> 'I used to be called Nick Chopper, when I was a woodman and cut down trees for a living.'

> 'Good gracious!' cried the Tin Woodman in astonishment. 'If you are Nick Chopper's Head, then you are Me—or I'm You—or—or—What relation are we, anyhow?'

Many of these stories and philosophies are based upon intuitions about matter that the Greek Atomists would immediately grasp. Yet in modern physics our atomism is even more radical than that of the Greeks. The contemporary update to atomist theory comes in several flavors.

For example, all fundamental particles of a given type—all electrons, or all photons, or all top quarks—are posited to be indistinguishable. This is a far stronger statement than simply saying they are 'alike'. When counting the number of distinct microscopic states of a physical system, if we exchange electron A with electron B, those two states (AB and BA) are counted as *the same microstate*.

This is very strange. Put differently, if we swap an atom on one side of a room with an identical atom on the other side of the room, it's as if the swap has never occurred. In fact, we can swap every atom in the room with every other atom, in all possible combinations, and we still must count that as only one distinct microstate. The count of microstates isn't changed by identical particle swaps. The collective microstate only changes if the position or velocity of one or more of the atoms is changed.

Now add to this mix the additional strangeness of quantum mechanics and dig even deeper into quantum field theory and the Standard Model. Here we find a roiling vacuum and sea of virtual particles, popping in and out of existence in a restless dance, the forces between particles emerging from the constant exchange of other particles…it all becomes a bit much for the old-style materialist mindset to take in.

All these theories are products of early or mid- to late-twentieth century physics, and they now rest upon a firm experimental foundation, confirmed by many tests over the decades. The more recent and even wilder theories of dark matter and dark energy, string theory and the multiverse, are still seeking confirmation. The Jorge Luis Borges' short story, 'The Garden of the Forking Paths' (1941), is a literary expression of the notion that any time we make a choice, to turn left or right, to step forward or back, that the world entire splits and continues along each separate path, while we and the world around us ramify, each fork now a stranger to one another. This short story is sometimes invoked to explicate the quantum multiverse in literary form. But for our purposes, it's more evocative to imagine the story of infinitely forking paths as told from the center of a turbulent maelstrom: in the quantum multiverse the forking doesn't depend upon our personal choices, but instead all the possible outcomes of each chance encounter of every fundamental particle.

The modern atomist revolution is still unfolding, and the radical character of 20th century physical theory has not fully penetrated the areas beyond the sciences—areas where some intuitions still seem to be based upon antiquated notions about matter, with its solid billiard ball atoms banging around, governed by cold equations, the future determined by the past, as if the universe was a giant clockwork where our fates are fixed at the outset of the world. This is a picture of things which physicists found they had to abandon over a century ago. Atoms are not billiard balls. The world is not a clockwork. In addition, light is not composed of particles, like Newton believed, nor is it a wave, like some 19th century physicists believed. Instead, light has the characteristics of both wave *and* particle. As do atoms, which are both localized and wavelike. If we insist on thinking of them as little billiard balls, we must endow them with a restless shimmer. Chaos is the order of the day, not determinism.

◦━◦━◦

Why does any of this matter outside of the physical sciences? Because our intuitions about the stuff the world is made of can pop up in surprising places, like political or religious ideologies—serving as a kind of metaphysical fog that clouds and creeps into places it shouldn't. Science, the shared and open enterprise to develop reliable knowledge about the world, can morph by degrees into the ideology of *scientism* by a certain type of move, in which sloppy, outmoded thinking is passed off as 'objective' and 'rigorous'.

The late physicist Freeman Dyson made this point eloquently in his essay collection *Infinite in All Directions* (1988). There he writes that he'd recently attended a talk by a famous biologist who spoke about there being two philosophical viewpoints which he called 'scientific materialism' and 'religious transcendentalism'.[2] At heart, the two world views were incompatible, according to the biologist. Dyson went on to say that he didn't understand the meaning of the word 'materialism'. To him, 'matter' in modern physics is an 'imprecise and rather old-fashioned concept' because in our current understanding of things, matter is how particles behave when large numbers of them are clumped together. But, when viewed in isolation, we see those same particles behaving 'as an active agent rather than an inert substance'. That roiling vacuum is what he has in mind here, not the void of the atomists but instead something far stranger, full of fluctuating virtualities. The actions of these particles in the strictest sense are unpredictable, they are not the billiard balls of 19th century materialist philosophies.

Dyson goes on in even more radical fashion to conclude that:

> Between matter and mind as we observe it in our own consciousness, there seems to be only a difference in degree but not in kind. If God is accessible to us, then his mind and ours likewise differ from each other only in degree and not in kind. We stand, in a manner of speaking, midway between the unpredictability of matter and the unpredictability of God.

Like Wittgenstein beating against the cage created by language in order say something he believes deeply, Dyson here is attempting to share his inmost thoughts using the imprecise language available to him—thoughts that were inspired by insights revealed to him in mathematical equations, written in a language that most of us cannot read. But those laws of physics are a place where consciousness might be hiding if we only understood the potentialities of what the equations are telling us. If we could only navigate their complexity, we might glimpse ourselves peering out of all those churning and unpredictable particles of which, it seems, we are made.

Dyson was one of the leading theoretical physicists of his generation. He was a co-creator of quantum electrodynamics (QED), the first theory that fully aligned quantum mechanics and special relativity, and the Director of the Institute for Advanced Study at Princeton for many years (where Einstein spent his last two decades). His views are his own, of course, and I suspect that his theological speculations are not widely shared among physicists, who mostly keep their thoughts on these matters to themselves. But the point Dyson is making is one that I believe most physicists would certainly agree with: the 'atomism' of modern physics is not the Ancient Greek atomism of Leucippus, Democritus, and Lucretius. The matter of modern physical theory is not the kind of matter that forms the metaphysical ground of 'materialism' as that word is generally understood outside physics. The materialist Emperor isn't wearing the clothes he thinks he's wearing, and his head is encrusted with the barnacles of long-dead physics.

Modern physics is far from finished. Neuroscience cannot yet explain consciousness. And AI is still far from creating a true machine companion. But all these fields are making progress. Great discoveries await. All these fields are undergoing sustained upheavals that will carry us well into the middle of the current century. Such developments counsel great humility regarding our old ways of thinking. If the aim of ideology is to eliminate ambiguity, as the writer Margaret Atwood has argued,[3] then we must be ready for the fall of ideological idols, because the very ground beneath our feet is shifting.

※※

When I was in high school, one of my teachers asked the class to reflect upon what we wanted to become in life, and to write it on a slip of paper. Without a moment's thought, I wrote: 'I want to be free,' and passed it forward. He shook his head when he read my note, as if world weary, and told the class that here was someone who hadn't yet woken up to the fact that no one is really free. Time to grow up. Instead, what I took from that day's lesson was to be careful who I shared my dreams with, and to hide my attempts to run against the bars of the cage.

With atomism we have become like the Tin Woodman—who, when confronted with his own self, realizes that the questions 'Who am I?' and 'What am I?' are fatefully entwined. And to this we should also add: 'Am I free?' The philosopher Jenann Ismael ponders the question of human freedom in her book *How Physics Makes Us Free* (2016). Ismael's point seems to be that the typical arguments invoked to claim that Newtonian physics, Einsteinian physics, or even quantum mechanics have no room for human freedom are not couched in clear terms. They often look to the character of the laws of nature, or to the mathematical equations that express them, and *voilà*—suddenly there is no room in them for freedom.

Ismael herself engages more seriously with the question of freedom within the physics of complex assemblages of matter and comes to a very different conclusion. That complexity of interacting particles can evolve to become what she calls 'self-regulating'. This does not mean they break out of the cage of physical laws, but rather that they exploit the properties of matter to create skeins of signal pathways and feedback loops. These loops can become so complex that, even with the constraints of physics, it's possible to develop what we can call a story-telling engine, matter speaking to itself—an internal 'I' that models an external world that the 'I' is situated within. That 'I' and 'world' model emerge from the hubbub of all that chaos that lies within us at the level of fundamental particles.

I find this idea intriguing, if not entirely convincing. The question of situating human freedom within the prevailing scientific world view is always something worth pondering, if only to try out certain new mental tools that we can use to beat against the walls of Wittgenstein's cage of language. Such give and take between science and reflection is essential: human freedom matters a great deal, even if we can't quite understand it, or fit it within what science tells us about the world.

～～

I once had a conversation with a colleague, a biologist, who told me that when the question of free will comes up in class he tells his students that they have no free will, only a perfect illusion of it cooked up by the brain. I asked him how he would ever prove that empirically, and he looked at me like I had spoken a heresy. But all I was really asking for was to show some humility. A dash, just here and there, to avoid getting stuck with old bromides that won't stand the test of future generations. Closing off certain lines of inquiry as either prematurely settled, or simply out-of-bounds, says to our students that some of the deepest and most meaningful things in their lives must be put aside when they enter a laboratory or a lecture hall. Instead, we should encourage them to keep beating against the walls of the world using whatever languages we have available, if only so we can send messages to one another, like signal fires in the night.

CHAPTER 9

Stirring the Ashes of our Dreams

We have lost our future and our past, our deep contact with nature, and the belief that we can truly know ourselves. Is it any wonder some feel a tug for the dreaming spires of the Middle Ages?

❧

On the evening of April 25, 2019, around 6:20 pm in Paris, reports began to circulate online that the cathedral of Notre Dame was in flames. The old timbers were dry, and the roof was quickly engulfed. For a time—before investigators determined the fire was an accident—social media was ablaze with conspiracy theories. The possibility that a tossed cigarette or some faulty wiring might have led to such extensive damage to the famous cathedral seemed implausible to many, a rejection fueled by the faulty intuition that big events must always have equally momentous causes.

Within a few days of the fire, hundreds of millions of dollars had been pledged to rebuild the cathedral, even as many other sites of worship around the world were besieged: churches put to the torch by haters, synagogues and mosques the sites of massacres.

Yet with pain and grief around us, many still felt that seeing Our Lady burn in the aftermath of an accident was a singular blow to the heart. Why did this event matter so much?

Most obviously, the cathedral was a visual icon: part of a shared cultural heritage, familiar from photos and movies, a backdrop for lovers across generations. A place for dreamers who strolled the banks of the Seine at sunset, or, still tipsy at dawn, came to watch the rising sun glint off the spires. For those with a more prosaic turn of mind, the medieval cathedral stood as evidence that civilization could endure through centuries of war, riot, and revolution. For the French of all faiths, or those of no faith, the cathedral was part of their national identity. Yet none of this seems to capture the full significance of the event.

The cathedral was built on the Île de la Cité in the 12th century, atop the ruins of two earlier churches, which themselves were built above an older Roman temple to Jupiter. The Christian Church attempted to bury earlier pagan or indigenous histories in this way, a pattern that repeats itself again and again, from York Minster in England to Monks Mound in Cahokia, Illinois. The dramatic central spire of Notre Dame which collapsed in spectacular fashion on the night of the fire was in fact a late addition, constructed only in 1859. Our Lady has seen the rise and fall of kings and empires, revolutions, and five republics so far. Angry mobs during the French Revolution dragged statues of kings from the cathedral out into the public square for beheading; they believed the sculptures represented kings of France, though in fact they were the ancient kings of Judea.[1] And still the flying buttresses and vaulted ceilings of Notre Dame itself remained intact down through the centuries. Those structures carried no threat to any ideology; perhaps they were recognized as feats of engineering, staking their claims against the disinterested hierarchies of gravity, rather than propping up human political systems.

When the fire broke out, I surprised myself by feeling compelled to share images of it in real time. Ordinarily I tend to lurk, an un-watched watcher of the human drama. My habit is to suspend judgement until the facts clarify; I am a scientist by habit of mind, not only by training. But on the night of the fire, I was moved, and wanted to express myself as a part of that larger trauma while it was ongoing. I began reposting images and videos. Why?

King Priam of Troy, upon hearing news of the death of his be-loved son Hector, fell to his knees in grief and, taking a handful of soil from the ground, poured the dust over his head. In the mod-ern world, while many are willing to share their most intimate thoughts and experiences on social media, others are more intro-verted, too self-conscious for such outward expressions of emo-tion. Many of the causes of our grief are too diffuse anyway, too ramifying and various, coming one upon the other. News of the death of loved ones might allow us to speak our personal grief, at least among those who share in it, but we can be less prone to do so in relation to the more distant news of the death of cultures, or even whole species. Psychologists now study various forms of climate anxiety, especially among the young, and they have even coined a new term for it: *doomerism*. There seems to be a well of unspoken grief we hold in common. In just the last few years, a global pandemic has killed millions, and genocidal warfare has returned to Europe. And in the US, normally calm observers of political and social trends now speak openly of their fears about a coming spasm of violence, or even the end of our experiment in democracy. Perhaps the flames of Notre Dame not only burned away old wood but turned to ash the old and comforting idea that, whatever comes, our civilization will surely endure.

Will our civilization endure? That question will not be an-swered by the longevity of buildings, but by our ability to form relationships that sustain us and give us courage to face what is coming. I recall a trip to Washington, DC, when I was about ten years old, and how proud my father was showing off the US Capitol to a visiting friend from England, Ann O'Connor. Ann had served in Whitehall during the Second World War, when it seemed to many that Western civilization was self-immolating. Ann had worked as a radio operator in the war rooms beneath Westminster and encountered Winston Churchill every night as he prowled about seeking news of the war. She recalled those years not as a time of fear, but almost winsomely, as a period in her life when she was young and people like her felt a shared sense of purpose, knowing they were fighting on the right side. She and my father first met at a dance in London, and that night he talked non-stop about my mother back home in New York City. My parents had only just married, and Ann was charmed by this lonely man, sensing his kindness and decency, and they became lifelong friends. She wanted to meet the woman he was so smitten with, so she and my mother soon began a chatty correspondence which lasted for almost fifty years until my mother's death. Like those who lived through dark times in past generations, we can choose to believe that after darkness there will come the light. We should make friends and dance while we can.

In her book *Hope and Grief in the Anthropocene* (2016), the Australian geographer Lesley Head argues that modern society suffers from two forms of grief. The first involves looking backward in time and is the grieving for a lost past. This is not just nostalgia for the world of our childhood, but paradoxically, the sense that we have lost a past that never really existed, the ground upon which human civilizations stake claims to national identities. Non-Western societies often have the comfort of creation myths that stabilize their understanding of origins, but the origin stories of nation states seem to shift in each retelling.

The pantheons and memorials that can be found in places like Washington DC, London, or Paris, attempt to situate the nation in eternity, to stabilize a sense of national identity. But the more we learn about our past, the more complicated the story becomes.

The Yale historian Timothy Snyder, in his course 'The Making of Modern Ukraine' available on YouTube, argues that a distinct Ukrainian identity is being forged on the battlefield and on the home front. He begins the course by telling something of the history of 'history' as an academic pursuit, its role in freeing us from cant and dogma. We can identify false histories, by contrast, based on their attempts to simplify and justify national myths. These are stories, for example, that imply that the nation has always been here by finding its 'true' origin in a deep connection to earlier idealized civilizations such as the Holy Roman Empire, Ancient Greece, or Rome. This makes the nation seem timeless and situates it as the inheritor of some imagined imperial grandeur or sacred purpose. But such myths overlook the fact that nation states are recent inventions in the larger story of human social evolution, and their true backstory is most often one of territorial appropriation and the cultural erasure of prior civilizations. The national myth is simple and heroic, but a truer sense of history, one that uncovers the full complexity of the past, destabilizes the comforting national myth. What once seemed a solid understanding of the past dissolves under the acid test of listening to other sides of the story, as told by those who were long silenced or ignored. Head's first form of grief, then, flows from the realization that we are descendants not of gods and heroes, but of some mix of a divine spark and mud.

The grief for a lost past is a legacy not only of the study of history, but also the advance of science. Historians refuse to let us believe the fairytales of the old national origin stories, and they reveal many national heroes to have feet of clay. Meanwhile, scientists tell us we are star stuff, begat of the lightning of distant stellar explosions, bewildered animals embedded within processes that span the galaxy and play out in deep time. Beyond that nugget of knowledge, science is silent on what it all means and agnostic on the question of purpose.

The second form of grief that Head identifies looks forward in time. It concerns the loss of a future we now realize is likely to be forever out of reach—a future of ease and steady progress, both social and technological; or perhaps a world where the science of social progress is perfected and comes to be seen as a kind of technology itself. Utopias always involve history coming to an end, the achievement of perfection. In our imaginations, the future can look like brightness trapped in amber. But we now know that climate change is only beginning, that the train has left the station and all our efforts to date amount to trying to take our foot off the accelerator. Even if we succeed in the transition to net-zero emissions by 2050, we have set things in motion that will require future generations to continue to adapt. Glaciers that provide fresh water for half the world's population are receding quickly. Seas are becoming more acidic because of the uptake of carbon dioxide from the atmosphere, preventing crustaceans from making proper exoskeletons. Ocean temperatures are rising, causing coral bleaching and the destruction of biodiversity. Global migrations of unprecedented scale are underway, not only of humans but also plants and animals and whole ecosystems. The trees of the world's temperate forests are already stressed by the requirement to migrate poleward, their seeds carried on the winds of change. Humans can adapt when circumstances require it, but can we do so when we carry all the world's biodiversity in the lifeboat with us?

Here, too, science says that a quiet utopian future can't be true, that complex systems like societies and ecologies never settle down, never reach an end state. They instead pass through a series of temporary local equilibria, where niches can sustain things for a time, before ruptures happen which upend the balance once more.

It's the ability to survive such tumult that makes living things so tenacious.

Looking back to our origins and forward to our possible futures, science always feels half-finished and iconoclastic. So be it. Bring on the artists and dreamers, tinkering in their studios and workshops, exploring the byways of the human spirit, to help us play and find a home in the world and, most importantly, to help us find meaning as we struggle to survive the cataclysms that are coming.

As things stand, humanity has yet to create a shared remembrance of the past by seeking out a truer version of our histories, and we have so far botched our imagined future because we did not really plan for it. We do not yet provide for our children's children by adopting open-ended ways of doing things, technologies and social forms that can be sustained for the long haul, as opposed to the next few years or decades. In our creative destruction, we are so often like a farmer who casts seed on stony ground and then walks away.

These two forms of grief, both forward and backward looking, according to Head, are left largely unremarked and usually suffered in silence. Until we come to grips with them, until we acknowledge and process them fully, we will not be able to move on to a more realistic stage of hope—the grim kind of hope held by one who is a dying, who sees things clearly, who knows that she is about to pass from this life, and yet still believes there is reason to act for the good of others who will live on after her. Until our societies achieve this level of steely-eyed realism about what we face in the coming decades, tempered by a deep and open-hearted love of the world, we will never truly deserve to put our name *Anthropos* on the current age.

Where Head argues that humans are burdened by two forms of grief, Hannah Arendt claims in *The Human Condition* (1958) that we suffer from two kinds of alienation. Many of us live cocooned in built spaces that separate us from the natural world, a familiar form of alienation for those dwelling in cities. But the modern sciences of the mind have also revealed that we are separated from our true selves and alienated from our own deepest desires. The discovery of the unconscious over a century ago was just the beginning; by now cognitive science and the neuroscience studies of perception have shown that so much of what the brain does happens below the level of conscious awareness, where sight and sound and other sensory signals are sorted into parcels of meaning. It's now understood by scientists that our brain creates a kind of story for our conscious mind about what's happening, rather than giving us direct access to the reality around us.

If we have lost our future and our past, as well as our deep contact with nature and the belief that we can truly know ourselves—is it any wonder some feel a tug for the dreaming spires of the Middle Ages? But those dreaming spires are also a story we've told ourselves, the idea that those centuries past were a deeper and more spiritual age than ours. Perhaps it's only because what's built in stone lasts longer than what's been fashioned from timber and dirt. In fact, careful historical investigation shows that many other construction projects from the Middle Ages were far more practical in scope: road building, canals, irrigation, waterworks, all made of wood or piles of dirt now decayed or merged back into what looks like a natural landscape to the untutored eye. Those more prosaic projects collectively consumed much labor and effort, perhaps as much as the cathedrals. But they are largely erased and forgotten, and in the erasure, a more spiritual age seems to emerge like the recollections of a dream.

Why don't we celebrate the workers who dig the canals that nourish the fields and feed the people—at least not as much as we venerate those who carved the sculptures and gargoyles that glower out over Paris all day long? Within a few days of the Notre Dame disaster, I began to read the usual claims that cathedrals were special marks of human civilization, that cathedrals were the only human endeavor that stretched across generations, in which a workman might enter the project long after its start, contribute his life's energy, and then pass on long before its completion. I appreciate the poetic imagery, but I respectfully disagree that this is something that distinguishes cathedral-building from any other form of human endeavor. It's not so much the focus on the multigenerational nature of it, which is generally lacking in our world of market-based transactions and just-in-time supply chains. What I object to is the claim to its uniqueness. Such claims separate us more than they bring us together; after all, there are other subcultures and forms of collective achievement where we still labor for the ages, so to speak.

The effort to create a working fusion reactor, for example, has been a significant multigenerational effort. I have known fusion scientists that have spent their entire career on that work and went into retirement knowing that a commercial reactor was still a generation or two away. This willingness to keep the faith, to know the reward lies somewhere over the horizon and will only be enjoyed by others, is true for most scientific research. Its practitioners see that science is always a work in progress and will never be finished. Beyond the sciences, my colleagues in the humanities teach their students to think about research and scholarship as joining a long conversation—one that started before we were born, and one that will continue long after we leave the scene. We enter that larger human conversation by first listening, then by joining the flow. And we can only hope that after we're gone, some of our ideas might linger in the minds of others.

I began this essay one evening a few weeks after the Notre Dame fire, and then put it aside, hoping to return to it after it had stewed a bit more and things had calmed down to allow more reflection. But life intervened, along with events in the world at large, including the 2020 US election season and the January 6 insurrection at the US Capitol.

That last event was like a torch lit to the home of American democracy. During that long-ago trip to DC with my family as a ten-year-old, we'd visited those same hallways and marbled spaces. We had managed to slip into the gallery of the Senate on a day of quiet decorum and hushed voices in the chamber. It was Ann O'Connor, my father's old friend from London, who talked our way into the senate gallery that day, even though we didn't have tickets. She could be cheeky and assertive, and a risk-taker. By letting her speak while we remained silent, she convinced the guard that we were a family from England on a tour of America. Below us on the senate floor, one senator spoke his mind to a nearly empty room. I don't remember what the speech was about. I later learned it was something called a 'colloquy', a highly stylized process that involved leading questions posed by a colleague, followed by elaborate and detailed answers.

What I learned that day was that democracy is often boring to watch. And yet, the building with its great dome and heroic statues, and the slow and plodding proceedings, seemed to radiate a solidity that spoke to me as if American democracy had always been here, and always would be. That feeling seemed unshakable to me as a child, and even later in life when I learned that the Capitol had been built by enslaved people, and that violence had sometimes come to the very floors of Congress during the run-up to the Civil War. All that messy history seemed a distant story about another world long past. But in the senate chamber where we listened to the colloquy, a half-century later the self-styled QAnon Shaman sat briefly in the chair reserved for the Vice President, rioters ranged the halls in an attempt to block the peaceful transfer of power, while others wiped feces on the walls.

We are living through what political scientists call a 'rupture', in which all the rules and power relations of geopolitics are in a state of flux. An energy crisis, a political crisis, a food crisis, and a financial crisis, all entangled with the climate crisis. All these dark trends threaten the idea that mere words can do much to make things better. But to believe that would be to let our grief at being abandoned on the road between past and future overwhelm us, to swallow us in isolating silence, rather than using it to remind us of what's at stake. We should instead try to articulate our sense of loss, to use it as a touchstone to provide moral clarity, because it also reminds us that some things are worth fighting for. And so, we stir the ashes of our dreams to see what might still be of use as winter comes.

CHAPTER 10:

Beware the Orwellian Trap

In which I examine how the evidence against a conspiracy can always be flipped in its favor, by appeal to the conspiracy itself. 'The conspiracy goes deeper than first imagined.'

❧

'What you're seeing and what you're reading is not what's happening.'
—DONALD TRUMP

'The party told you to reject the evidence of your eyes and ears. It was their final, most essential command.'
—GEORGE ORWELL

'Before they seize power and establish a world according to their doctrines, totalitarian movements conjure up a lying world of consistency which is more adequate to the needs of the human mind than reality itself.'
—HANNAH ARENDT

In his book *1984* (1949), George Orwell's most famous work, the climax of the story comes when the protagonist Winston Smith is tortured until he comes to believe whatever Big Brother tells him is true, even if it contradicts the evidence of his own senses. Extreme ideologies, like the totalitarian one depicted in *1984*, can set traps by claiming to have an explanation for everything, ensnaring us into believing things that are untrue. The trap works by exploiting the fact that none of us perceives reality directly; an objective understanding of things must come through shared experience and the creation of a common language to describe it.[1]

There is some resonance with broader politics here. Political movements are largely conjured from the dreams and fears of people who are otherwise strangers to one another. Politics at its best allows large numbers of those strangers to cooperate and work toward common goals. At its worst, it can turn us against one another. The reasoning traps that can occur in politics, the totalizing ideologies that style themselves as all-encompassing world theories, or the paranoid conspiracy theories that cannot be proven wrong—all of these were a preoccupation of Orwell's throughout his life. His attempts to navigate a path around them can be found threading their way into his novels, essays, and even his newspaper columns written during the Second World War, where he constantly called out the ways in which political speech and even the meaning of words could morph from one day to the next to suit the needs of those in power. It was as if he viewed modern life as consisting of one long struggle against being gaslighted. Hence his constant repeated admonitions to use clear language and avoid dehumanizing abstractions.

Orwell's fixation with how we construct a shared political understanding through language was likely inspired, at least in part, by his participation in the Spanish Civil War and the siege of Barcelona. As recounted in *Homage to Catalonia* (1938), he witnessed certain events in combat that were then falsified in later narratives tailored to serve a political purpose: to turn the leftist factions against one another to secure the Stalinist party's total control over the one true story of 'what happened'. Repeated, again and again, the Stalinist party line gradually took over the general understanding of things, until others who took part in those same events that Orwell witnessed came to believe it. An entirely new reality was created to serve political purposes. Orwell refused to accept the new reality and had to flee for his life.

If you find yourself in such an Orwellian world, where up is down and freedom is slavery, it's almost impossible to escape. When an entire body politic finds itself in such a pass, we might feel like an animal with its paw in a trap, forced to chew off a limb in order to escape despite the risk of a later infection. But the real question is: how do we avoid the trap in the first place, and how can we escape from it whole once caught?

We live in a time when most scientists agree that humans have accelerated the progression of climate change, yet the citizenry of many nations isn't fully convinced. Those who think the skepticism has been fomented by our crop of current leaders—and will cease when they step away from the scene—are mistaken. Once a significant fraction of the population begins to believe a separate set of theories about reality, to tell a different and internally self-consistent set of stories, it becomes very difficult to come back together. That will take hard work, open minds, and a dose of humility all around.

If being part of a culture includes sharing intuitions about how the world works, then those of us alive today—even living in the developed world, which might seem homogenized by modernity—are not part of a single technological culture. Even within individual countries we see a disordered collection of interpenetrating subcultures, each with different intuitions and beliefs about the way things work.

For example, pedagogical research in the 1980s into what makes the study of physics difficult for many students found that it was partly because of many students' pre-formed and incorrect intuitions about the world. Quizzes revealed that many believed that the natural state of things was one of rest, or that rotational and linear motion were fundamentally different, both central tenets of Aristotelianism. To learn Newtonian physics, one must first unlearn these prior intuitions, which can be very hard work. Those unfortunates who are never graced with the joy of taking an introductory physics class might never unlearn those intuitions, and so wander the earth using a kind of 'folk theory' of the world supplanted centuries ago by scientists.

Many students also hold intuitions about the nature of visual perception that are simply wrong, involving unseen rays that leave the eye and illuminate the scene. This active theory of visual perception is called 'extramission'. It was inherited from the Ancient Greeks and survived well into the Middle Ages.[2] The developmental psychologist Piaget reported that young children seem to believe in extramission, and at least one recent study found that up to half of current college students still believe in it, apparently because they never unlearned it from childhood.[3] It remained the best available theory of visual perception until around the time of Kepler, when his more passive theory of ray optics, called 'intromission', was successfully used to design telescopes and microscopes. This passive theory holds that the eye merely takes in the rays that arrive from a scene that has been illuminated by other light sources. No rays are emitted by the eye. Yet the intuitive appeal of extramission, with its poetic sense that we might somehow grace a loved one with our gaze such that they feel it as a gentle form of touch, ensures it remains in circulation.

Clearly, we inhabit different countries of the mind even as we peacefully mingle at the local coffee shop or chat around the dining room table with friends and loved ones. But in times of great change, we risk being pulled apart if these differing world views become polarizing.

But wasn't it ever thus? What's distinct about the present is that the public square now intrudes into the private sphere through our screens, pummeling us with obfuscations, misrepresentations, elisions, bullshit, gaslighting and outright lies. We are living through a Cambrian explosion of online subcultures, some with incommensurate world views, many of which are highly politicized. The volume of these lies and their often-visual character is unprecedented, tailored for us by algorithm. Soon after the internet slipped from the narrow confines of scientific collaboration, porn, religion, and conspiracies were its most popular sites. By now, our personal hopes, fears, and secret desires have long been commodified and monetized. Those sites of seduction aren't yet Orwellian traps, because they don't necessarily distort our shared sense of reality. But they do make the setting of traps much easier by breeding a culture of distrust and cynicism, by atomizing us from one another by amplifying our differences, and by subverting the hope that there might also be nobility or honor in the public sphere.

It's easy to understand why people lie in politics. Some are willing to put personal gain over the public good, while for others it's a power move, an assertion that they 'create' the reality the rest of us must live within. These are not errors in reasoning but defects in character. The viral particles of distrust that these lies have set in motion make the body politic sick at heart. The only long-term cure is to find stronger antibodies against liars, even as Orwellian traps have hijacked that part of our cognitive immune system.

The political scientist Tom Nichols has argued that much of our current toxic politics in the west, including the tendency to believe in conspiracies, might be due to the boredom of living in a relatively affluent society.[4] This 'itch for chaos' theme, and the dangers of widespread boredom, can also be found in the writings of thinkers as diverse as George Steiner, Eric Hoffer and Bertrand Russell. Perhaps the conspiracy simply makes the world seem more *interesting*.

But more than outright lies, or even alternate facts and propaganda, what interests me here are alternative theories that carry their own explanatory power: totalizing ideologies and global conspiracy theories that people sincerely believe as they try to puzzle out a complex and dangerous world. Such theories are attractive because they help the world make sense. Internally, they are fairly self-consistent as narratives and tend to have a deterministic character, with causal chains working in lockstep according to some secret or devious plan. The idea that the world might in fact be fundamentally chaotic, with governments and social institutions held together only by extraordinary human effort—such a fragile existence is more frightening than believing the world is run by a malevolent but highly competent star chamber. Big 'shit' doesn't just happen; someone makes it happen. If the person or cadre responsible for world events isn't immediately apparent, then they must be hidden, acting behind the scenes.

To be clear, sometimes major catastrophes have clear human causes, and the list of suspects can be quite short. It's true that Russian president Vladimir Putin and his inner circle are responsible for the recent invasion of Ukraine. But the fact that they have not been able to prevail in conquering the Ukrainian people just shows how limited Putin's power really is. The kind of conspiracy theories of interest here are instead global, secret and all-powerful.

One fact that must be faced is that the appeal of conspiracy theories flows from our love of stories. Our brains are hardwired to organize experience into narratives, to spin tales as a way of making sense of things, and our need for stories is both emotional and intellectual.[5] My colleague Deborah Morse, a scholar of Victorian literature, once told me that the wide interest in spiritualism in the late-19th century could be understood as a response to a yearning that grew out of the absence of God, a way to fill a void that the advance of science had opened in the human heart. So perhaps we should also see the enduring appeal of conspiracy theories as reflecting the ancient human need for gods and demons as the hidden agents acting behind the scenes, as a way to tell their stories in an acceptably modern guise. The emotional appeal of conspiracies might simply reflect a rejection of the idea that we live in a chaotic world without intrinsic meaning, one where shared meaning must be created. Instead, those prone to conspiracies perhaps believe that surely someone must be pulling the strings, stirring the pot to create all the chaos of the modern world.

Social scientists coined the term 'conspirituality' about a decade ago, when they noticed two streams of thought merging online.[6] The first was the current version of New Age spirituality, which hews to notions, inter alia, that we are evolving toward a higher plane of being, that there are unseen dimensions beyond what science knows about, and that alien or angelic races with higher levels of consciousness and power are watching us. All have in common the assertion that there is more to the world than our current science can comprehend. The second stream is a paranoid mix of ideas that casts the world as one controlled by a secret cabal that really runs things. Here we find the usual list of suspects: the Jews, the Rothschilds, the Freemasons, Satan, the CIA, the Trilateral Commission, the Illuminati, the Deep State, the Lizard Overlords, George Soros. From this list, it's clear that such paranoid theories live on both the Right and Left ends of the political spectrum, and among the religious and the nominally secular. This reflects what political theorists call the 'horseshoe' character of politics, where rather than lying on a simple linear spectrum of left and right, extremes on both ends embrace overlapping conspiracies, leading to the possibility of a grand ideological synthesis based on quackery.[7] Conspirituality concerns the merging of these two strains of theorizing about the world New Age spirituality miscegenating with paranoid politics, with QAnon being the latest variant.

Ideology and conspiracy theorizing thus go hand in hand here, because if your ideological theory of how the world really works keeps getting proved wrong, one possible explanation is that there is a secret conspiracy of the powerful working against it. Conspiracy-style theorizing can be more dangerous than lies, because evolution has gifted us with an inborn aversion to liars, a mental gag reflex. But we have scant immunity to paranoia. Instead of lies, the preferred bait for Orwellian traps is the seduction of certitude. Therefore, the most dangerous conspiracy theories of all are the ones that can't be proven wrong.

※～～※

In some sense we are all living in a dream, even those of us who believe science is our most reliable guide to finding out how the world works. We lack direct access to the reality around us; instead, our brains use the noisy and faulty data of our senses and our memories to construct a rough theory of reality from moment to moment. Most of this goes on below the level of conscious awareness, so in a sense we are all sleepwalking. Only by adopting a shared language, and a commitment to truth-telling, can objective knowledge of the world emerge.

Once a conspiratorial mind-set is adopted, though, the usual rules of evidence and objectivity are abandoned, even if some simulacrum of reasoning is retained. Reasoning becomes a kind of performance without real meaning, because the openness to surprise and a willingness to admit we might be wrong have been frozen out.

I teach my students that there are always an infinite number of theories consistent with any given set of data. How then, can we ever use observations to improve our understanding? Philosophers call this the problem of induction, and it's what prevents our theories of the physical world from ever being as certain as the truths of mathematics. We build our understanding of external reality using the evidence of the senses, and yet those senses are faulty. So, we must proceed with caution, and with an extra dose of humility about the limits of our knowledge.

The first step is to recognize that inducing something from observational data always relies on prior knowledge. New observations are interpreted through the filter of everything we already understood to be true about the world.

To make this more concrete: suppose I have a cousin, Rachel, who wins the lottery. The chances of her winning were one in six million, so she collects a big prize. That's great news. Now suppose I have another cousin, Vinnie, who goes to the casino and plays the dice, rolling sevens ten times in a row. This run of luck also has a likelihood of roughly one in six million. Will Vinnie be taking home a huge payoff? More likely he'll be thrown out of the casino for cheating.

Both outcomes carry the same odds. So why are we more willing to believe that Rachel's win is honest, but Vinnie's is not? Because we have prior knowledge that some people cheat and cheating at dice by palming a loaded pair is much easier than cheating the entire lottery system. The outcomes of the two gamblers' trials are considered 'new data'. But our instincts for how difficult cheating would be in each case are part of our 'prior knowledge'.

'They cheated' or 'they made it all up' is one of many possible theories that can explain any data set. 'God made it that way' is another familiar explanatory tactic, used by harried parents to shut down the questioning of curious children. Variations on these explanatory tropes are endless, and they can sometimes be quite subtle and convincing. One such theory in circulation now is that what we take to be the physical world might really be a computer simulation. This is a modern variant of Plato's Cave, where those who have only ever known life in the cave, watching shadows on the wall, mistake these shadows for reality. But unless we can leave the virtual reality cave—unless there is some way to tunnel our way out of our level of reality to reveal the underlying hardware running the simulation—there's not much to say about it. It's like Chuang Tzu, a philosopher of ancient China, who went to sleep one night and dreamed he was a butterfly, but upon awakening he asked himself whether he was a man who had dreamed of being a butterfly, or a butterfly now dreaming he was a man?

There are explanatory theories that cannot be proven wrong through observation, because they are consistent with all possible observations. A theory that can't be proven wrong, however, isn't automatically right. Nor is it automatically wrong. Instead, the truth of the theory is *empirically undecidable*. To be useful beyond providing the cold comfort of certitude, there must be a real possibility that our theory of the world can fail the test of observation.

❧

Consider now what is called the 'climate consensus', captured most completely in the accumulating and increasingly alarming series of UN Intergovernmental Panel on Climate Change (IPCC) reports. These reports are written by large committees of experts who try to summarize the current best understanding as it appears in the scientific literature, updated every few years. As someone who once took part in a similar exercise for the US

fusion program, I can attest that such documents tend to be con-
servative in their conclusions. Wild-eyed alarmist manifestoes
tend to be written by individuals, not large committees.

The skeptic's counterargument to the climate consensus goes
like this: science isn't a democracy. Scientists don't vote on the
truth or falsity of scientific theories. True scientists are led to their
conclusions through independent study of the evidence. So far, so
good, though it's a caricature of the complicated process by which
observations and theories come to be accepted or rejected. But
let's go with it for the moment.

To be fair, the history of science is full of examples where the
consensus view turned out to be wrong. Some of these cases are
what we'd now call 'scientific revolutions': the germ theory of disease, relativity, evolution, quantum mechanics. These were not
merely ignored or considered incorrect for a time but were believed to be *unacceptable* given the then-current understanding of
what a good scientific theory should look like. So, let's acknowledge that we must remain open to surprise, and that sometimes
those surprises overturn a reigning consensus. Fair enough.

The social media meme 'consensus isn't science' is correct as
far as it goes. But taken as a nugget of a thought, 'consensus isn't
science' is merely a small seed. If we are to judge a seed by the tree
that grows from it, this one never seems to sprout. It promotes
reflex responses, not considered opinions about how science
should inform policy.

'Consensus isn't science' is a truism, but it's largely irrele-
vant to any discussion about what we should do in response to
a changing climate. It deflects from the real question: among all
the competing explanations and theories, which of them presents
the wisest guide for action? We don't have the luxury of hopping
into a time machine to find out which of the current competing
theories will turn out to be correct, and which will be left by the
wayside. We must choose between theories *right now.*

Imagine you are driving with friends on a dark and foggy night, and you become lost. The only person in the car with a map says they think a bridge has collapsed up ahead. What should the driver do? Step on the gas? Keep going at full speed? Or slow down? This is how we ought to think about the climate crisis, according to the late Cambridge physicist David J. C. McKay in his masterful book *Sustainable Energy—Without the Hot Air* (2008). We can understand it more generally, however, as a metaphor for avoiding Orwellian traps. If our map reader believes she's spied a trap ahead, it's best to slow down and proceed with caution.

In conversations I've had with climate skeptics, a subtle shift sometimes occurs at a critical point in the argument. Instead of saying that we should always keep in mind that the consensus *might* be wrong, the argument becomes: we can be assured the consensus is wrong *precisely because it's a consensus*. This is very strange. An equivalent argument would go like this: suppose you are in a room with no windows, and you wonder if it's raining outside. You ask a hundred people who have recently been outside and 97 out of 100 tell you it's raining. Why would that be taken as convincing evidence that the sun is shining? After all, in science there is a much higher than 97 percent consensus that atoms exist, that the speed of light is a fundamental constant, and that the Moon is not made of cheese.

Stripped of its political salience, the error in reasoning against the climate consensus is obvious. Why then do so many climate skeptics fall into this trap, even those who do not invoke 'consensus isn't science' as a sterile rhetorical move? Put differently, what are the alternative theories that might lead someone to reject a scientific consensus regarding climate change?

There are several. One is that climate scientists are all radical anti-capitalists. But if we buy into this theory, we'd also have to include the editorial teams at the flagship newspapers of capitalism: *The Financial Times, Bloomberg, The Economist*. All have come out strongly in support of science-based policymaking that would guide us in the transition to a carbon-neutral world. It can hardly be the case that every one of them is a member of a conspiracy to bring down capitalism.

Another theory is that climate scientists are lemmings who simply follow the senior leadership in their fields, even when they are being carried off a political cliff. Or that climate scientists are somehow self-hypnotized, entrapped in a group delusion, a massive form of Orwellian groupthink, massaging their data to support their prior conclusions.

Sadly, the history of science shows that all these scenarios have occurred at one time or another, at one lab or university or another, and they cannot be assigned a zero probability in the current case. Every human system is flawed in design and botched in execution. But the questions are always: how flawed? How botched? And where should the wise place some measure of confidence? If you are forced to bet your family's future, how should you bet *right now*?

✦

A few years ago, I attended a talk by the legal scholar Dan Kahan of Yale, on science communication failures. Like many of my scientist colleagues, I had assumed things like climate change denial and vaccine hesitancy were due to a lack of scientific knowledge. If this is true, more science education should solve the problem. But Kahan's research findings were eye-opening. There was little correlation between climate denialism and scientific expertise. In fact, if you measured someone's general scientific knowledge ahead of time, and then plotted how they were likely to respond to a new question about human-caused climate change, you could see a clear divergence into two camps: between those who believe it's happening and those who doubt it. And the greater the general scientific knowledge, the *stronger* the divergence of opinion. In other words, those who doubt climate science the most are sometimes extremely well informed about science generally.[8] So what's going on?

Kahan and co-workers at Yale have shown that for most policy debates, most of the time, people come down in alignment with what science would suggest is the correct course of action. But there's a small subset of science topics which he calls 'pathological'—things such as climate change, vaccinations and gun violence. In these areas, political identity trumps science. That is, if a piece of evidence threatens your political identity, or undermines your ideology, the listener's tendency is to either treat it with a higher degree of skepticism or ignore it entirely.

There's reason to be hopeful, however. Not all scientific matters touch on political identity. In any case, Kahan's later research suggests that one way to overcome the problem of reactively rejecting scientific evidence is to appeal to people's natural curiosity about how the world works, that childlike sense of wonder that we can all feel. That is something where we can find common ground and begin to talk to one another with respect. Another finding was that people were more open to taking in bad climate news if the article was accompanied by a summary of actions that could be taken to make things better, rather than being left only with a sense of dread that things will inevitably get worse.

What's needed, at a minimum, is an admission of the fact that sometimes science communication fails. If we are to make progress addressing things like climate change, we need to try and understand how to keep those subjects off the list of pathological science topics. We need to do a better job of understanding how people consume information and use it to form judgments of how the world works, and we need to help people learn to avoid Orwellian traps.

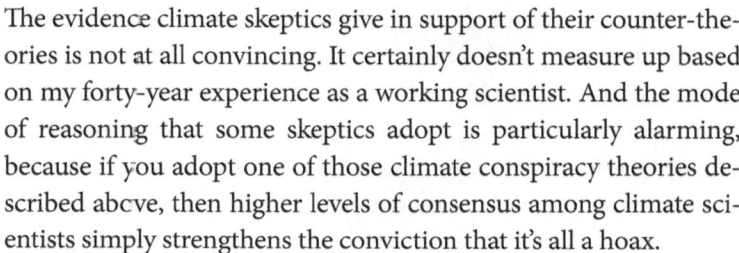

The evidence climate skeptics give in support of their counter-theories is not at all convincing. It certainly doesn't measure up based on my forty-year experience as a working scientist. And the mode of reasoning that some skeptics adopt is particularly alarming, because if you adopt one of those climate conspiracy theories described above, then higher levels of consensus among climate scientists simply strengthens the conviction that it's all a hoax.

The Orwellian trap has been sprung. Evidence against a conspiracy can always be flipped in its favor, by appeal to the conspiracy itself. *The conspiracy goes deeper than first imagined.* The adoption of such a far-ranging and ever-inflating conspiracy theory creates an information filter that drives you in one direction: towards paranoia. Therefore, we have to recognize Orwellian traps for what they are, so we can avoid putting our foot in them, and help others avoid them, too.

But how can you get out of an Orwellian trap once it's snapped shut? Gnawing a leg off to escape it, if we stick with the body politic as metaphor, would entail abandoning many of our fellow citizens. But that cannot be the right answer. Some skeptics are led to adopt a conspiracy mindset because of an understandable terror of being cut off, of being left behind, of being abandoned in an ever more dangerous world. For others, the skepticism of what experts tell them flows from childhood trauma, where trusted adults betrayed or abused them. Research has shown that anti-vaxxers are far more likely to have experienced abuse or abandonment in childhood than those who support vaccination. That fear of great change can be real, and the distrust of authority figures can be based on personal life experience, even if belief in a conspiracy is a false antidote.

We can also have compassion for the coal miner or oil worker who will be thrown out of work during the coming transition to a zero-carbon world. Some of them, perhaps, have been seduced into believing that climate change is a hoax in hopes that they might be able to continue to work in a job they know, in a community they love, and where they hoped to raise a family. The green transition should be carried out with these vulnerable communities in mind, to create new opportunities for them in the emerging renewable energy economy. And we should condemn those who seek to profit politically from their pain.

In the long run, the only way to get out of the Orwellian trap as a collective is to use our whole being, our hearts and minds—not to mention raw cunning—to figure out how to un-spring the trap and overcome our fear, to create bonds of trust, and help others do the same. The current age of conspiracies reflects a loss of trust in governments and our institutions. We need to keep talking, and listening, while also moving forward at speed—reminding ourselves that the best strategy for winning another's trust is to first be worthy of it.

CHAPTER 11:

Gaia Shrugged

Our experience of the weather is true, but our knowledge of the climate is objective. Why is that distinction so important?

～～

Standing cn the shore in the summer of 2016 at Grand Isle, Louisiana, I am struck by the magnitude and ineffability of what is happening to the Earth's climate. Grand Isle is a barrier island at the outermost edge of the Mississippi Delta, overlooking a wide, beautiful beach that has been artificially replenished with fine white sand. Off-shore in the distance the rust-red outlines of a few oil platforms dot the horizon, pumping crude for delivery to the refineries that lie to the north, while closer in a shrimp boat plies the waters just beyond the surf, nets spread out in the afternoon sun. Everything is peaceful today, yet everything is in motion.

Where I stand at Grand Isle, solid rock lies far beneath my feet. Over millions of years, the Mississippi Delta added more and more sediment as the interior of the continent eroded, layer after layer, flood after flood— a process that formed a compacting coating of silt so thick that bedrock is only a theoretical concept here. The porosity of these surface layers, combined with rising sea-levels and the resulting erosion, means that the Mississippi Delta region is losing roughly one square mile of dry land per week.[1] A northward migration of people inland is underway, a retreat from the advancing sea: generation by generation, it will likely continue for the rest of this century. For people affected by erosion, from the coastal US to Bangladesh and the Marshall Islands, global sea-level rise is personal.

Weather is something we can perceive directly through our senses. It is often immediate and undeniable. We can feel rain and snow on our skin or taste it on our tongue. We can see the effects of the wind as it tosses the branches of the tree beside us. Climate, however, is an abstraction. It is the average temperature, or mean high tide, or average rainfall. It is the shifting baseline behind those daily weather events. We experience weather, but we can only observe the climate. Weather is what we see when we look out the window. Climate is the notional baseline about which things fluctuate.

~~~

Just what is the average sea level, for example? This is itself an abstract concept, an idealization shared and measured by instruments arrayed around the world, riding atop the waves, underwater, and on satellites overhead. If the Earth were made entirely of water, isolated and motionless in space, without currents or seamounts to complicate things, it would form a perfect sphere, held together by its own gravity. But once it's set spinning, centrifugal effects cause it to bulge outward at the equator relative to the axis of rotation, forming what is called an oblate spheroid. Now add the Moon in its orbit about the Earth, and the tidal forces it exerts raise up another bulge. If the Earth were not rotating, the tidal bulge would align with the Moon, but the Earth's rotation carries the tidal bulge forward a bit, the line of symmetry moving slightly ahead of the Moon's position in the sky.

If that were the entire story, the shape of the Earth would be an ellipsoid. Sea levels are locally measured relative to such an imaginary 'datum' surface, not a sphere. In principle you could use the sphere as the datum, but then global measured sea levels would vary by as much as thirteen miles. Using the ellipsoid as the reference reduces that variation globally, relative to the datum, to about a hundred meters. In practice the local datum is a 19-year average of tide data, tied to the orbital variation of the Moon of that period. Now add the land, that bumpy and rugged covering which has its own fractal shape. The rotating Earth, turning over once a day, drags the land surfaces beneath the tidal bulge which follows the Moon in its monthly and more stately motion. Therefore, the bulge sweeps around the Earth, at the equator traveling at roughly a thousand miles an hour, generating all those familiar flows in bays and bights and tidal wetlands.

The TOPEX/Poseidon spacecraft, launched in 1992 and operational until 2006, circled the Earth once every ninety minutes. TOPEX gave us the eyes we needed to see the true shape of the world's ocean. While the Earth rotated beneath it, the satellite scanned a swath of ocean with radar along a track roughly three miles wide, building up strips of data that had to be patched together and merged to form a global picture. Every ten days TOPEX accumulated a global snapshot of average sea surface height. Over the months and years of the TOPEX mission, and that of the follow-on satellites named JASON, scientists accumulated an enormous record of the shape of the Earth's oceans, measuring average sea surface height at each point to within a few centimeters. At that precision, you could squint through the ups and downs of waves and see instead the local baseline of the average sea surface. And with it you could now observe the track of major ocean currents, such as the Gulf Stream. The visualization of data from TOPEX made the oceans look like they were cut through by major highways and drifting vortex patches, moving transport belts that carry heat, nutrients, and living things around what otherwise looks to the unaided eye at sea level like a trackless ocean. The gravitational pull of seamounts hidden far beneath the surface become easy to pick out as well, dimples on the ocean's average face, along with the sloshing of the equatorial Pacific associated with phenomena such as El Niños, oscillations called Rossby waves, and other slow-moving patches of current. TOPEX and JASON allowed us to visualize for the first time the shape of what Melville called the great tide-beating heart of the planet.

❧

Climate concerns longer spans of time and greater distances than we can see with our own eyes. We can only detect the climate signal through collective rule-based observation, augmenting our senses with technology. This is how we developed our remarkable ability to see the tangled web of feedback loops and those global flows of mass and energy that keep the world in motion. While the 1957 novel *Atlas Shrugged* by Ayn Rand eventually became the urtext of ideological individualism, the name of Gaia, grandmother of Atlas, was adopted in the 1970s by the environmental scientists James Lovelock and Lynn Margulis, who saw climate and life as mutually co-evolving. We require the testimony of others to bring such things into view. Yet this is also why humanity's knowledge of climate is objective while our personal experience of weather is not. This distinction can be puzzling until we dig into the question of how scientists come to know things about the world, and the nature of objective knowledge.[2]

The key difference between an experience and an observation is that the former is forever singular and irreducible, while the latter uses systematic methods that means it can be repeated by another, at least in principle. This allows an observation to become a commodity of sorts, transportable and detached from personalities. An experience is always true because it is experienced, but it only becomes an observation when it can be shared and re-experienced by another.

The writer Amitav Ghosh makes a similar point in his book *The Great Derangement: Climate Change and the Unthinkable* (2016). Ghosh argues that it's no coincidence that the novel appeared as a literary form at about the same time as the emergence of science: both the novelist and the scientist must grasp rule-based watching, accompanied by a shared language used to describe what is seen. The novel is a literature of observation.

Objective knowledge, then, is not truly achievable in isolation. Instead, it must be constructed. It lies in the spaces among us, built upon and understood through an agreed-upon set of rules for observing, coupled with a vocabulary that allows those experiences to be compared across time and space.

Consider again the case of sea-level rise. The US National Oceanic and Atmospheric Administration (NOAA) says that of all the sites along the US East Coast and the Gulf Coast, the tidal gauges at Grand Isle, Louisiana, are registering the most rapid change in relative sea-level. These gauges are attached to docks or piers, and the sensor readouts are collected every few minutes and then made available online.[3] The network of tide gauges form a ring around the coastal US, and many others are deployed worldwide; they constitute an international pool of knowledge, from which an attempt is then made by scientists to create an objective understanding of the situation.[4]

These tidal gauges, along with satellite data, allow NOAA scientists to estimate that sea-levels globally are now rising on average about three millimeters a year, about the width of one of the folds in the skin on your knuckles. This change is due to a combination of the warming of ocean waters, which causes them to expand, and the addition of outflow from melting ice that resides on land, in glaciers and ice sheets. (If the ice is already floating in the water, then melting it doesn't change the global sea-level.)

Adding to that average rise in global average sea-level, local relative sea-level can change due to the land subsiding, or changes in coastal ocean currents, since the average sea surface height is depressed in the presence of those flows. At Grand Isle, the land is sinking twice as fast as the seas are rising, giving a total change in relative sea-level of about nine millimeters a year. That's almost the length of the last bone in your index finger. Every year, sea-level rise results in a steady encroachment of salt and brackish tidal waters into groundwater and freshwater streams, driving the migration of species and turning vast areas that were once oak forest into salt marsh, interspersed with dead tree trunks that look like desperate hands reaching up toward the sky. And the loss of land through erosion also leads to cities and towns further inland losing their protective buffers, making them more exposed to storm surge during hurricanes.

Trying to glimpse the future of sea-level rise in Grand Isle means we need to follow the state of what are known as the *grounding lines* of glaciers on the far side of the planet, in Greenland and Antarctica. These are essentially naturally formed dikes holding back a river of ice that would otherwise flow much faster into the sea. If those grounding lines give way, the rate of increase in sea-levels would accelerate dramatically. The fate of our coastal cities is therefore determined partly by what goes on a continent or more away on the dark sea bottom, in frigid waters kilometers below the surface, as a result of skirmishes between rock, gravel, sea, and ice.

Around the globe, underwater autonomous robots sample waters, acoustic sensor arrays measure the ocean's temperature profile, and humans dive to see the state of things with their own eyes. Elsewhere, they walk on Antarctic ice sheets and Arctic tundra, or fly in carefully mapped transects across the Greenland ice pack to measure its shape and thickness, to talk to one another, to tap data into machines, and to mine our growing midden pile of data.[5] This is an international conversation, the largest and most complex collaboration in human history, aimed at trying to puzzle out the objective state of the Earth's climate in order to make Gaia's tangled web visible—to fit all that data to idealized models and turn those terabyte arrays of numbers into human knowledge, an image of reality in the collective mind's eye of humanity.

Humanity's fate therefore depends upon events that are happening, or will happen, at places that only a very few of us will ever see with our eyes or stand upon with our feet. Many of these places will in fact be monitored by our machines who will gather the data for humans to analyze. We must trust others to act as our sentinels, to keep an eye on them for us, to use those technological extensions of our senses to help us see in those dark places, and to make visible the invisible, because if those frozen stores of water give way the world will quickly turn wild indeed.

Anxiety about the future, as well as an abiding sense of alienation and loneliness, are aspects of the modern human experience. Like pre-literate humans awakening to their own mortality, we are awakening, perhaps not for the first time, to the mortality of our civilization. The good news is that humans are an adaptable species, and inventive. Yet this genius for invention is precisely what has brought us to our current pass. We created a global civilization but based it upon the unsustainable foundation of fossil fuels.

Historians writing in some distant future will (hopefully) see fossil fuels as an easy path taken early in our technological development, pursued before a better one came into view. There is reason for cautious and guarded optimism about this prospect. Surveys suggest that most people understand what's at stake for the Earth's climate system and want our political leaders to do the right thing. But it will not be easy, and it will not happen quickly. It will be the work of generations.

If it's true that a culture ultimately lives in the brains of those who choose to take part in it, then the same is true of our shared sense of objective reality. The modern scientific notion that only the collective can achieve objectivity, through the creation of systems of shared subjectivity, points to a central vulnerability in human societies. Without a shared reality that allows us to map individual thoughts and experiences, one onto the other, we risk more than just a breakdown of political systems. The very bonds that allow us to build civilizations in the first place threaten to come apart.

For all its faults, and they are legion, modern civilization allows many of us to spend our days doing more than growing food or hunting, scrounging for a night's shelter, or fighting off predators and people trying to steal our food. This achievement also means that we must trust strangers to do the right thing most of the time, and to tell the truth as they best understand it, most of the time. To act in good faith, most of the time. Every system has its cheaters, but a system that doesn't believe in the possibility of a shared reality can't last very long because the non-human physical world—ever present, inescapable, and not remotely cognizant of human desires or wellbeing—will eventually overwhelm us.

≈≈ ≈≈

Five years after my visit, Grand Isle suffered a direct hit from Hurricane Ida in August 2021. It was one of the most destructive storms in Louisiana history, second only to Hurricane Katrina, and damaged nearly every building on the island.

When I was there, I found Grand Isle to be a welcoming village—a working town with a school, a library, a little league baseball field, and a community center, the civic buildings all up on stilts. Human beings are resilient, and our societies are adaptable. On the day of my visit, Our Lady of the Isle Catholic Church proclaimed on its roadside sign: 'I believe'. Yet the church was lofted well above ground and therefore just a little closer to heaven, not taking chances.

On the dune break at Grand Isle, next to one of the numerous signs that declared 'No motorized vehicles!', I stepped aside for an elderly gentleman in his motorized golf cart. He sputtered forward up the hill, lugging his beach umbrellas, canvas chair, and drinks cooler, out onto the wide strand proper; a determined human being, head down, either not noticing the signs or choosing to ignore them. It's then I noticed a half-dozen golf carts already parked along the beach, scofflaws every one. The scene was dotted with colorful umbrellas and a smattering of sunseekers. I waved at them, and they waved back. It was early June, before school was out, and the beach was not crowded at all. In fact, it was nearly empty, a quiet, calm day.

CHAPTER 12:

Can We Create a
Climate for Hope?

*How can we sail against the headwinds that are slowing progress
on climate change? It helps to remind ourselves that there are
countercurrents beneath the surface.*

~~~~~~

The city of Istanbul lies astride one end of the Bosporus, a deep and
narrow neck of water that flows in two directions at the same time.
On the surface, freshwater streams southward from the Black Sea towards the Mediterranean, while the denser salt water of the Mediterranean forms a northward countercurrent far beneath the surface. If
this hidden salty watercourse flowed overland, it would be the sixth
largest river in the world.[1] The Ancient Greeks knew of it and would
hitch a ride by dropping large sacks filled with rocks, exploiting the
Bosporus' hidden momentum to carry them past the crushing rocks
believed to have threatened Jason and the Argonauts.[2]

Adapting to climate change, and mitigating its worst effects,
will also require navigating dimly perceived currents of social

and political transformation. We need to do nothing short of re-inventing human civilization. The thought is terrifying, at least until we remind ourselves that each generation reinvents what comes before it. Climate change isn't a problem that can be fixed in one or two generations, let alone one or two election cycles. Instead, climate change is now, always has been, and always will be humanity's constant companion. What's different today is that the observed pace of change has accelerated, and the acceleration is due to human activity.

The climate news can be depressing. Attempting to sail against the headwinds that can stymie action on climate change, it's help-ful to remind ourselves that there are always countercurrents be-neath the surface, pushing in the other direction. By riding those subsurface streams, we might still get where we need to go.

While rigid ideologies can dictate policies in the short-term, they tend to become gradually less relevant as the world changes around them. Liberal democracies should have a greater capacity for reinvention and renewal than some other political systems, since they are not created in the belief that there is a single, doc-trinally correct solution to any given problem. At least in theo-ry, democracies should be well suited to respond to the evolving challenges presented by climate change, so long as citizens sup-port policies that will make a big difference over the long haul. But we can't realistically expect those same citizens to sustain a sense of urgency over the generational time scales on which cli-mate change unfolds. What to do?

❧

Computers and computation are overused as tropes when ap-plied to humans and society. But they're apt insofar as they en-courage us to think about what we might call the 'default set-tings' for our civilization's 'operating system'. In a computer, the system defaults guide the decision paths the computer follows when it boots up. Change one default setting, and your background changes color. Change another, and you have a different pre-selected Wi-Fi login. Because of advances in technology and changes in social mores, each generation effectively 're-boots' our civilization. So, if we want to understand what strategic moves we should make to put ourselves on a more humane and resilient footing, one that is more benign and has a better chance of lasting centuries, we should attend to the default settings for society's operating system.

Some examples of defaults are well known: a carbon tax, properly applied, would be a kind of default reset to economics, as would the rapidly emerging renewable energy economy. In some markets, solar and wind are already the cheapest form of power for utilities. Smart grids, improvements in batteries and other energy storage technologies will also dramatically reset the defaults for how we generate and distribute power in coming decades. The mix of government-funded research and development, tax incentives, and direct subsidies that encourage commercialization, have been shown to deliver huge returns that benefit everyone.

Other kinds of default settings are less obvious. Take the 'greening' of forensic accounting, financial risk analysis and insurance underwriting. This would mean insisting upon rigorous audits of companies to understand how investors and insurance companies are exposed to environmental risks—not only from the effects of climate change such as increased risk of fires, storms, and floods, but because of regulatory risk in the event of carbon taxes, and market risk due to rapid advances in renewable energy. If a company has no meaningful climate risk strategy, investors should take their money elsewhere, and insurance companies should decline coverage. This would enforce market discipline.

Resetting defaults also involves things like creating new cur-
ricula for the civil engineers and architects who will design, build,
and maintain the infrastructure we will need in the coming decades. Much of the basic infrastructure of our major cities, the
subway tunnels, bridges, and water supply networks, were put in
place in the late 19th and early 20th centuries. When entering
New York City by car or train, you most likely travel on bridges
or through tunnels first built before the Second World War. This
suggests that many major infrastructure projects built in the next
few decades will (hopefully) still be in use in the year 2100. By
then, sea levels along the East Coast of the US are projected to be
several feet higher than present, and the rise is accelerating. Rainfall patterns will change across the globe, and with them flood
and fire zones will shift. Civil engineering and urban planning
students will need to know how to design for a changing world.

If life expectancies continue to increase along current trends,
then a good number of my current university students will still
be alive at the dawn of the 22nd century. Yet a recent survey of
almost a thousand civil engineering students revealed that half
do not believe in anthropogenic (i.e., human influenced) climate
change. These are the future designers and builders of our infrastructure, charged with helping us to become more resilient, and
to help the next generations survive and flourish.

Resetting defaults also means upgrading design guidance
and codes for buildings, homes, and facilities. In 2018 Gover-
nor of Virginia, Ralph Northam, signed Executive Order 24 requiring all state agencies to do a climate resilience audit, and
to upgrade design requirements on all state buildings using the
best available climate projections. While things have been slow-
er at the federal level in the US, strategic planning for climate
adaptation is moving forward in those regions already hit by sea
level rise, such as Hampton Roads, Southeast Florida and the
Mississippi Delta. These actions are based upon people's direct

experience of climate change, happening in the here and now, not in some ill-defined future that affects someone else.

There are yet deeper defaults that persist at the level of our political philosophies, as well as our legal systems and government institutions. The philosopher Stephen M. Gardiner has argued that climate change is fundamentally an ethical problem, and that it presents us with a 'perfect moral storm': an intergenerational, interspecies, and truly global quandary. These are areas where historically the Western tradition has had serious blind spots.[3]

For example, the Stanford Dictionary of Philosophy entry for 'intergenerational justice' notes that, even though philosophical discussions of what we owe to future generations date back to antiquity, John Rawls' *A Theory of Justice* (1971) was the first attempt to consider that question within a systematic theory of justice. As climate change calls all our basic assumptions into question, it seems unconscionable that generations long dead can limit our actions through wills, testaments, entails and covenants, and that current generations can bring claims before the courts, but future generations have no legal standing whatsoever currently recognized by US courts.

It's also becoming harder to defend the idea that corporations can be legal persons, while non-human beings that are capable of suffering, and ecosystems that sustain human wellbeing, cannot sustain harm as far as the law is concerned. Instead, they are mere resources to be utilized for the benefit of legal persons. Under the Western legal tradition, a legal person has rights, can own property, and can sue if they are harmed by another person. A legal non-person has no rights, can be owned as property, and has no standing to sue no matter how egregious the damage they suffer.

The wall separating legal persons and legal non-persons may be starting to crumble. Indigenous cultures often find the Western notion that the non-human living world is a mere resource for exploitation to be a repugnant idea. Some of the changes underway can be understood as part of the global Indigenous Rights movement, which is helping Western institutions and laws negotiate a new relationship with older traditions. For example, the Whanganui River in New Zealand, which is sacred to the Māori, is now classed as a legal person, as is the Ganges in India. In 2008 Ecuador became the first country to recognize what are called the Rights of Nature, which means that entire ecosystems can be considered legal persons, which has already had an impact on mining and logging industries. The Costa Rican city of Curridabat has granted citizenship to pollinators and trees, ensuring that their wellbeing must also be considered in town planning.[4] Efforts to expand legal personhood to animals like elephants, gorillas and chimpanzees have so far failed in the US courts, but have met with success elsewhere. Future courts might be called upon to recognize that not only future generations but the biosphere itself can have legal claims on us, and their wellbeing must be considered when now-living humans make their plans. Taken together, these examples suggest that our civilization is at least beginning to reinvent itself and to rethink its basic ethical assumptions in light of the climate crisis. Such changes expand the scope of our compassion and concern, so they include not only our descendants, but also a larger part of the natural world.

This work, from improving our ability to manage the financial risks of climate change, to upgrading guidance for construction, to a foundational shift in our laws, courts, and government institutions charged with protecting our wellbeing—these can all push us in the right direction, one step at a time. That's provided we keep the longer-term goal in view: the creation of a far more livable and humane civilization, that has a much lighter footprint on the wider natural world.

Human society is not perfectible, but it is improvable. Such work can be deeply unsettling because it calls into question and forces us to reflect upon our deepest values. Ultimately it can also

be a source of hope, precisely because it taps into the creative energies of our entire society to meet the climate crisis. Collective
endeavors allow us to ride the deeper counter currents that are always present, even when stiff winds are blowing us the other way.

CHAPTER 13:

Hooked on a Feeling

How the humble Big Mac helped to bring down a mighty empire.

❦

Denial is a form of self-protection, but it can also be accompanied by a kind of catastrophic determinism. That's how voters might believe that truly dangerous, planet-imperiling activities would surely be banned by the government or might think that the world is doomed to burn because of the venality and weakness of our leaders. At times they might believe both contradictory stories to be true.

Such dissonance can lead to paralysis, especially when politicians fear voters will punish them for taking decisive action to prevent climate catastrophe. That doesn't mean that individuals can't make reasonable, evidence-based decisions in their everyday lives. What's worrying is that a collective paralysis emerges at the larger economic and political level, where the wisdom of crowds is supposed to operate. There, where the community should pull us toward a more positive future, we're often stymied by a belief in what psychologists call an 'external locus of control'.

I first encountered this concept while reading Lisa Margonelli's book *Oil on the Brain: Petroleum's Long, Strange Trip to Your Tank* (2007). A masterful storyteller, Margonelli starts by recounting how a trip to buy gas at her local Kwik Mart led her to a sudden realization that she had little sense of where gasoline came from, so she set off on a years-long quest that took her around the world to try and understand the oil industry. Pulling on that thread, she dove into the history of the petroleum industry, arranged visits to a regional wholesale distributor, a refinery, an oil field, the NYMEX futures exchange, Venezuela, Nigeria, Chad, and Iran—all with an eye towards understanding the world oil markets and how it gets from the ground to our pump.

At each step of the way, Margonelli found that everyone she interviewed—consumers, industry executives, government policymakers, wildcatters and speculators—all acknowledged that oil is bad for the environment, and bad for the many of the places in the world that are rich in petroleum and poor in good governance. They understood that oil wealth and corruption have gone hand in hand. Yet it always seemed to be someone else's fault. Consumers said that the fat cats of Big Oil are fouling the planet for personal gain. Oil executives blamed the consumer. Politicians in many places saw oil as a necessary evil, a transitional source of wealth they can use to develop otherwise poor economies. The locus of control always lay somewhere over the horizon, where those who 'really' had power remained anonymous and faceless.

The sense of powerlessness felt by good people entangled in bad systems gets in the way of meaningful action. We are never completely helpless. Those of us who live in democratic societies can take part in the political process, and we have some power in the marketplace by making informed choices about the environmental impacts of the things we buy and the food we eat. The big moves will have to be undertaken by governments and industry, but the ethical question of who should be held responsible for the current mess, and who should take the lead in getting us out of it, is one of extraordinary complexity. Here we might even talk about there being an external locus of moral responsibility. It's always the other guy who's at fault.

We can calculate the contributions to global emissions in various ways. For example, we might consider current emitters versus legacy emitters. The first group is dominated by China and India, the latter by the old colonial powers in Europe and the US. By either measure, the Global North dominates the Global South. Alternatively, we might look at how emissions are generated by income class, cutting across national boundaries, and find that the emissions of the rich of the world far outstrip the poor. This is true even within national borders, anywhere in the world. Or we might look at emissions by companies and find that the cumulative emissions from the start of the industrial revolution are dominated by around a hundred enterprises, a mix of private fossil energy corporations and state-owned entities.

Traditional theories of ethics are overwhelmed by the complexity of the problem of climate change, precisely because responsibility for it is so widely distributed and multivarious. At the same time, the question of who can make an impact is also a dauntingly difficult one. There are billions of actors who can only make a small contribution individually but who could collectively turn the tide if they moved in the right direction, while there's a much smaller list of actors, including government and industry leaders, who could make a much larger individual impact. It's easy for any one of these actors, big or small, to buy into the external locus of control. The consumer can argue that it's something the government and industry should do; politicians and industry leaders can argue that it's up to the voters and consumers. So, we are caught in a tangle, a wicked ethical mess, but one where giving up and bemoaning our lack of agency is as much a flaw in reasoning as an ethical lapse. As Rabbi Abraham Joshua Heschel wrote: 'Few are guilty, but all are responsible.'[1]

The fact is that *homo sapiens* are simply not very good at using reason as a guide to action. We are not so much rational beings as rationalizing ones, making up tales after the fact to explain why we do things. We most often work with intuitive rules of thumb based on small sample sets from daily experience or things we've heard from others, but we are not usually 'reasoning'—at least not in the sense that Enlightenment philosophers would use the term, meaning the careful use of rigorous logical analysis, taking care to seek out all relevant data and to consider counter arguments. The idea that our brains can think 'fast' by using such heuristics, while rational thinking is 'slow', has become mainstream in recent decades.[2] I first encountered it in the books by the American neurophysiologist William Calvin, who argued that the brain is constantly creating new narratives, sifting, evaluating, adopting or rejecting them in a never-ending churn, always trying to make sense of things. This idea also appeared in the writings of philosophers like Daniel Dennett and Patricia Churchland, which drew heavily upon the recent advances in cognitive science.

In *Reason in a Dark Time* (2014), the philosopher Dale Jamieson argues that the distinction between how our brains actually work, and the Enlightenment ideal of what it means to reason rationally, helps to explain our failure so far to address the threat of climate change.[3] This is because rational knowledge doesn't usually produce a change in our collective behavior. A social species, we look around us for clues to how we should act; so, if everyone else seems to be going about their day as if everything is in order, we carry on. But when we feel an electric urgency in our gut, we can act quickly, drawing upon a fight-or-flight response that kept our ancestors alive. For most of our evolutionary history, rough-and-ready decision making, rather than ratiocination, would have made the difference between life and death. Those who opted for a time-consuming, logical approach were likely to get eaten.

The prefrontal cortex of *Homo Sapiens* is a late arrival on the evolutionary scene, and it is not a logician but a storyteller. It spins tales for us about why we do things, helping us to make sense of a world that our ancestors would have found bizarre. Many of us are immersed in a techno-civilization that few, if any, of us can really grasp in its totality, and yet our storytelling can turn it into a home. The neocortex is also the part of our brain that allows us to imagine the future, even when we find that future dark.

Jamieson identified a number of key failures in our response to climate change—failures in our political systems, our economic theories, our public policy and planning, and in our efforts at science communication. These are all areas where rationality is highly prized; the abode of supposedly dispassionate experts who traffic in abstractions, streams of data and elaborate theories. Though they work outside my own field of research in physics, I recognize my intellectual kin. But many people feel disconnected from these expert communities, receiving messages and reports from them as if they were signals sent by an alien species speaking a strange language, trying to warn of an impending disaster.

Jamieson is right about our failures. Yet a dire situation is not the same as a hopeless one. Jamieson is not arguing that we should give up, but rather that we should see our prior failures as a call to dig deeper, to seek new ideas, to pursue new ways forward. Addressing climate change is the largest and most complex collective action problem in human history—but it's worth remembering that our ancestors, even those just a few generations past, would marvel at what we've already been able to achieve in other domains.

———

If the wisdom of crowds can lead to paralysis, and if reasoning is not what drives societies to act, what about our aesthetic and emotional responses? Are they shifting over time in ways that might be galvanizing? A shift in the things we find beautiful or ugly can be potentially significant. For example, the ancestors of many Westerners feared the wilderness. The *Oxford English Dictionary* shows that the word first carried negative connotations as a place of desolation, disordered, a region where one can easily get lost, or an uncultivated region not yet brought under the plow. By contrast, many now believe the physical wilderness can be beautiful in its own right and worthy of protection. A change in how we feel may also give cause for hope, especially when it's affirmed by a more 'objective' understanding of things. Xuebin Zhang, the oceanographer and research scientist who was the coordinating lead author of the most recent Intergovernmental Panel on Climate Change (IPCC) Report, says: 'Climate change is happening, and people actually feel it. The report just provides scientific validation to the general public that, yes, what you feel is actually true.'

In a 2020 New Yorker essay,[4] the science fiction writer Kim Stanley Robinson argues that the Covid pandemic catalyzed a major shift in public sentiment. Robinson relates how he embarked on a rafting trip through the Grand Canyon in early March of 2020, knowing that a novel coronavirus was already circulating in the US. When he climbed back out of the canyon a week later, it was as if he'd stepped into another world. Whole states and nations were in lockdown; cemeteries were overflowing in Iran, Italy, and Ecuador; reefer trucks used as makeshift morgues lined the streets outside some hospitals in New York City. One friend of mine from Brooklyn knew forty-five people who died that spring. As I write this in the fall of 2022, some parts of the world are groping fitfully toward full recovery, while others are still living in the shadow of a nightmare. It continues to loom over many of us precisely because of a collective failure to prioritize the health of other human beings, both through the underfunding of public health systems, and because of personal behaviors that rank individual freedom as more important than taking responsibility for avoiding harm to others.

~~~

As a newborn, I had to be rushed back to hospital soon after coming home because of a severe allergic reaction. My parents feared they would lose me. We lived on a farm at the time, surrounded by dogs and cats and all manner of plant and fungal allergens, and tests showed my allergies were triggered by almost everything. My earliest memories include weekly visits to the doctor for shots. Doctor Österreicher was a kindly man who would warn me of the coming jab, then praise my bravery when I didn't cry. I didn't feel brave, however, and while I hated the shots, I loved my doctor because I knew he wanted to keep me safe.

Even with the inoculations, I suffered from bad asthma attacks from earliest childhood until my teenage years. Those who suffer from them know they can be terrifying. Not being able to breathe triggers those parts of the brain that fear immediate extinction. That was a time before EpiPens and nebulizers, so we relied on some syrupy medicine my mother would give me with a spoon. Usually within half an hour or so I could feel the attack break, my breathing would clear up, and then I could rest. After the attack subsided, there followed an oceanic calm and some of the deepest sleeps I've ever experienced.

My father had also been asthmatic as a child. He grew out of it, like I eventually would. But then he developed the problem again later in life, after we'd left the farm and moved to Baltimore. Those attacks were probably triggered by exposure to air-borne pollutants at the shipyard where he worked, and after he moved on from that job the attacks never recurred. He'd also lost a crew mate to asthma, when he was a young man working onboard a ship in the 1930s. They'd been loading grain, and the floating cloud of aerosolized wheat dusting the ship provoked the man's attack and he went to his cabin to lay down. My father was the one who found him, dead, blue from being starved of oxygen. That could have been my father, in another time or place. So, my father learned not to follow his instincts during an attack, when panic drives us to ground, urging us to find a place to huddle. That's the primitive brain stem leading us down the wrong path. When the body fails, your greatest hope is in the company of others who might help you.

These memories came to the fore for me in the fall of 2021 as I read about the fourth Covid wave, which many feared would turn out to be the worst one yet. That rise in case numbers was especially tragic because it could have been avoided. I speak now primarily of the US experience, where we let our COVID response get caught up in the culture war, and where some think it's acceptable to use children's bodies as a battleground. Pathological public health behavior knows no political boundaries, however. Similar events are playing out in other parts of the world too.

Many forms of toxic politics around the world seem to be driven by a yearning to return to a normal that no longer exists. Such narratives can nudge our desires toward the familiar and secure, and away from the new and unsettling. But when the world is upended, the familiar might no longer be the safe harbor it seems. The dogged pursuit of our former life becomes misguided, if not deranged—the equivalent of acting as if a virus can be beaten down by sheer force of will, instead of careful consideration of the science.

Given the importance of vaccines to public health, those who embrace vaccines have shown a great deal of anger toward those who are hesitant or refuse to get vaccinated. I can understand that anger and confess that I've felt it myself at times. But my views shifted slightly after reading a recent study of the motivations of anti-vaxxers and those who are vaccine hesitant. The research found that this type of behavior tends to correlate with child-hood trauma, including sexual abuse, battering, an acrimonious divorce between parents, or abandonment.[5] Is it any wonder that trauma in childhood might engender a deep distrust of authority late in life? Our values—not the ones we espouse but the ones we actually live by—are an amalgam of things we are taught grow-ing up, attitudes that rub off on us from the society in which we live and work, and our personal experiences. This explains why acknowledging the human diversity of beliefs, hopes, dreams and fears can never be ignored when crafting public policy. To do so is folly.

We are all living and working in societies that are in deep de-nial about the state of things, looking away and refusing to face things squarely, as if we are walking on thin ice over a deep well of unarticulated grief. So perhaps it's best to see antivax senti-ments or climate change denialism as emotional responses that flow from an upwelling of grief over the end of dreams of safety, of imagined futures gone forever.

If some widespread shift in feeling is underway, driven by wit-nessing the events of the last few years, this could change our politics in ways we can't predict. For good and bad, old ways of thinking will come to seem outdated, and new actions could be-come possible that were unthinkable before the pandemic.

In the 1970s and 80s, behind the Iron Curtain, another shift in feeling had taken hold. Most of us in the US knew very little about it, though eventually it would bring an end to the Soviet Union. When I visited the USSR in the 1990s, I was struck by the dysfunction of the society around me—how everyone I met under the age of forty wanted to move to California, while those over forty expressed cynicism about the future. There seemed to be a complete dearth of optimism, a universal embrace of the idea of an external locus of control, fed by a desire to avoid whatever might be coming next. The television in my conference accommodation showed documentaries consisting of fact-finding trips to Western Europe and the US. Between MTV music videos, breathless news reporters gobbled fast food on camera and asked: 'Why can't *we* make a decent hamburger?'

The 2003 German film *Good Bye Lenin!* captures this shift in feeling with wit and insight. A woman who went into a coma prior to the fall of the Berlin Wall wakes up after the event. Her son tries to hide this fact from her, to protect her from knowing that the socialist paradise she had dedicated her life to creating had vanished while she slept. A shift in the objects of human desire had undermined the revolution—slowly at first, then quickly.

They couldn't deliver french fries worth eating, or popular music worth listening to. In the waning years of the Cold War, Soviet functionaries were viewed as an ossified, aging, and out-of-touch elite. The worm of seductive individualism had undermined the great Soviet experiment down to its roots, and the rot was now exposed. Collapse soon followed because, in the end, the state could not deliver what the younger generation wanted. I remember a smartly dressed Kremlin spokesman interviewed on US television, opining in flawless English: 'As Marx wrote, "From each according to his abilities, and to each according to his place in society."' No student of political theory, even I knew this was a corruption of the *Communist Manifesto*, cloaking a justification for voracious greed in garbled revolutionary rhetoric. But Marxist theory was no substitute for a Big Mac. The dubious claims to have met the abstract goals of 5-year plans for the beet harvest could not compete with images of Madonna dancing like a virgin.

Disruptive changes in politics and popular culture have a similar character, in that they often begin by a testing of boundaries, pushing against a door they believed to be closed only to find the way ahead was open all along. The Yale historian Marci Shore talks about such a moment during the Maidan protests of 2014 in Ukraine.[6] These started out as a protest by young students against the government's reactionary turn toward Russia and away from Europe. A future within the EU that had beckoned, with the possibility of study at European universities, jobs and life prospects abroad, things that upwardly mobile and well-educated young Ukrainians might aspire to—all this had been taken away from them overnight by government fiat. In their anger, they came out to Maidan Square in central Kyiv to protest. The government set the security forces upon them, beating them, hoping to terrorize not only the young but their parents, who would force them to stay home. Instead, at a critical moment, the parents joined their children out of a sense of moral outrage. That's when the protest quickly morphed into what came to be called the 'Revolution of Dignity'.

Shore says that the word dignity here should be understood in the Kantian sense. Anything that can be traded for another thing essentially has a finite price. That which cannot be traded is beyond price and only such things, like human beings, Kant believed, have true dignity. Shore watched the protests remotely through live streams, but even from afar she could tell this revolution was different. Most, she argues, are 'Oedipal' revolutions, a blow by the young against the old who refuse to give up power. But the Revolution of Dignity was cross-generational, about shared values and what it would mean to be Ukrainian in the future. In her later interviews with people who were in the Maidan, many spoke of how their sense of time itself had shifted during the protest. It was as if they lived in a separate bubble of reality, and they became far more aware that the present was a kind of membrane that separates the past, which is unchangeable, from the future, which is unknowable. In that moment, with that clarity of mind, new possibilities emerged. They rattled a door handle, and found it was unlocked after all.

In the late 1980s I taught a course about nuclear weapons, and I asked my students: do you think the Cold War will ever end? None of them could imagine such a thing. The question itself seemed unworthy of consideration. Ever present in the background of our lives until then, the Cold War ordered our conception of the geopolitical world. Yet a few months later the Berlin Wall came down. For a few months, young people paraded through the streets of Prague and Budapest, singing John Lennon's *Imagine*, believing a new age was coming into being, one of great hope born of new freedoms.

And then the world turned, history didn't end, the struggle for the human future evolved into other forms of battle—online, across social networks, in market relations. Political scientists talk about 'ruptures' that can occur about once a generation, when the young—for whom the last generation's history is just a story—start to move into positions of power and influence. The older generation, for whom the past isn't really the past, must eventually give way to the young. The old must come to accept that many of their ideas are out of date, but they also sense an estrangement from new ways of feeling, new desires they find unsettling, new forms of human emancipation they cannot accept. These emotion-driven shifts beneath our feet are just as momentous as new technologies and scientific understandings, that stuff of the rational mind. This is true because what we feel and what we find beautiful or ugly might, in fact, be what finally motivates us to act.

Where are the walls now about to fall, following a revolution of dignity? And how will we exploit these fresh breaches to build a better world? Which old ideas, old ways of doing things, old systems and institutions no longer fit for purpose—which of them will fade or evolve into something new, to put not only humans but large parts of the living world beyond price?

CHAPTER 14:

We Need New Stories of the 21st Century University

If our civilization survives the coming upheavals, universities will too, even if their form and function changes—because universities are what makes our civilization possible.

≈≈

'The book carries us, in a manner, into company; and unites the two greatest and purest pleasures of human life, study and society.'
—DAVID HUME, *DIALOGUES CONCERNING NATURAL RELIGION*

Do a Google search that begins with: 'academia is'. The autocomplete function, peering into the global hive-mind, came up with the following answers in late 2022: academia is 'broken', 'toxic', 'dying', 'pointless', 'not a meritocracy', 'abusive', or 'a cult'. And 'academia is killing me.' What has happened to the dreaming spires?

The trench warfare of campus politics, the budget battles, the internal insurgencies, the tendency of faculty to circle the wagons and shoot inwards, the paucity of jobs—these mutually reinforcing trends feed the sense that something is terribly wrong. In such an environment, we can readily forget what we are fighting for, and why it truly matters.

With their mission to preserve, transmit and enlarge human knowledge, universities should find themselves working at the nexus of social transformation, as engines, beneficiaries, and sometimes victims, of that change. It's therefore puzzling that so many discussions about the future of the university slip into narrow and practical concerns, diagnosing broken funding models, and focusing on workforce needs to the exclusion of long-term student wellbeing. If we talk only of value propositions and return on investment, of lifetime earnings and payback, we risk missing what's really at stake.

Universities can sometimes be difficult to love, yet their ideal continues to inspire. We need a new and better vision of the university, one more in tune with its current reality. Yet we also need a picture that remains faithful to the deeper values of the academy, the values that are worthy of love. As the philosopher David Hume argued, the shared pursuit of learning can unite two of the purest pleasures of life: study and the company of others.

The writer Robert Bringhurst has argued that poetry is what we do when we think deeply, breaking through the crust of the familiar to touch a more fluid layer of thoughts hidden below. If those thoughts emerge as words, they become what is traditionally understood as a poem. But our thoughts can manifest in many other ways, too. An invention can be thought of as a 'maker-poem'. A new theory of space and time might be a 'physics-poem', while a theorem can be a 'math-poem'.

I like this notion of deep-thinking-as-poetry because it encourages us to avoid distracting surface differences. Instead, we can come to see the artist and scientist, the thinker and the maker, the teacher and student, as engaged in the same vital species-wide project: the quest to know ourselves better, so as to find our way in a dangerous world.

Universities have been a part of that human social project for nearly a millennium. They have survived wars and famines, revolutions and plagues. They are among the longest-lived of human institutions and tend to be far more robust than governments or business enterprises. In his book *The Uses of the University* (1963), University of California Chancellor Clark Kerr noted that of the Western institutions in continuous existence since 1500, only about 80 survive. These include the Catholic Church; the parliaments of England, Iceland, and the Isle of Man; a very few family-owned companies; and over 70 universities. This remarkable longevity alone should give us pause when politicians and business leaders confidently declare that they know best what universities need to do and be. It's highly likely that universities will change in form and design in coming decades, and that we might even invent new kinds of universities. Still, they will almost certainly survive current social upheavals, assuming our civilization itself survives them, precisely because the university's mission to preserve, transmit, and enlarge human knowledge is what makes civilization possible.

But rather than cling to any single expression of this aspiration, it is the aspiration itself that we should fight to preserve. And we must acknowledge that this dream draws its strength from a very deep and universal human yearning to learn from, and with, other people. Productive rumination about the future of the university means not only struggling with the important matter of the economics of higher education, and the wider social impacts of university programs. It also requires that we remain alive to the *poetics* of the university: attending to how we see the university when we call it up in our imaginations; how we think about its role in the world; the stories we tell of it; how we imagine its past and future. And, most importantly, how we dream about the kinds of lives that universities make possible.

The history of the university is not a single story, and there is no reason to believe the future of the university will be simple either. There is also no single, true measure against which all institutions should be judged. Instead, we ought to see that modern universities are complex and prone to change, and that they are a disorderly collection of programs and schools that evolve and adapt. Despite clickbait claims that we face a future with only ten universities, there is no reason to believe we are converging toward a single model, or toward global uniformity, and certainly no reason to think that this would be a good thing.

Although it might be argued that the tendency of capitalism and market economics pushes everything toward commodified uniformity, this is a transient phenomenon when viewed within a longer time frame. Such a trend toward homogeneity among universities would violate all the norms of evolutionary processes, which in fact militate against monocultures and instead tend toward highly diverse ecosystems. Monocultures are fundamentally unstable, less resilient in the face of shocks from changing environments. It's the nonuniform, complex, and manifold nature of universities that explains their remarkable longevity. Precisely because universities are organic entities, governed by laws of variation and selection, they are also full of puzzling inherited traits. Like all living things, they embody their history. They are not designed so much as adapted to the current moment by way of contingent and often haphazard processes. The university is older than capitalism and Marxism, and neoliberalism is merely middle aged; it's a good bet the evolving and hybridizing university will outlive them all.

In an interdisciplinary course I used to teach with several colleagues called 'The Idea of the University', we asked our students to go through the exercise of drafting a charter for their ideal institution. We encouraged them to think about why their university existed, what role they wanted it to play in the world, who got to govern it, and how decisions would be made. We asked: what would you do differently? What would your ideal university value most? Would it be a global entity, living online, fully embracing a virtualized future? Or would it harken back to the original academy, set in a grove of ancient trees, close to the earth, sea, and sky, a place of physical presence and long conversations, surrounded by nature? Would your ideal university focus on serving inner cities or rural towns, indigenous communities, or isolated island nations, providing access to students who would otherwise never dream of going to university? Or would you focus primarily on charting a course into a more traditional academic stratosphere, dedicated to the creation and preservation of arcane knowledge, a true ivory tower?

Our students collectively wanted all these models represented among future universities. In a world as prone to disruptive change as ours, a world of uneven development and even more uneven wealth and privilege, there is no reason why we should limit our imaginations to one stereotype of the university in our 21st century stories. We should not restrict our imaginings to sylvan quads or cloistered courtyards, tree-lined walks or pebbled paths, theatre seating for large lecture classes where professors drone and students doze. We need new and better stories about the 21st century university, stories and poems that feed our dreams about what's possible.

The hunger for knowledge is universal, and as parts of the world long left behind shake off the lingering impact of colonialism, more and more people are pursuing education. According to the website *Our World in Data,* only 12 percent of the world's population could read and write in 1820. Today, that is roughly the fraction of people who remain illiterate. Universal literacy is no longer a dream, it is a near-term reality. Generation upon generation, as levels of education increase, so too does the demand for more advanced instruction. So, while we might think that innovative university designs will be largely technological—the online university, the university that's fully virtual—it's also important to remember that those technologies are merely enablers of new social formations and driven by local needs and desires.

For example, during the colonial era, education in many of the more remote islands of the South Pacific was still rudimentary—focused on learning how to read and write sufficiently well to serve on colonial plantations, and to read the Bible. But after the Second World War, and with the coming of self-determination, several newly formed island nations set out to create educational systems for their citizens. To do so, they had to invent a new kind of university. The University of the South Pacific was founded in 1968 as a collaboration of twelve island nations with a total land area about that of Denmark. It now serves a population of over a million people spread over 33 million square miles of the Pacific—a sea surface area larger than the entire Eurasian landmass.

There is also the University of the West Indies (UWI), based in Jamaica and founded in 1948. The UWI is a collaboration among the 17 English-speaking island nations of the Caribbean. When I met the Jamaican novelist Erna Brodber on a visit to my university, she told me that as a student of history at UWI in the early 1960s, she learned a lot about Scottish history but nothing about Jamaican history. So, after her PhD, instead of pursuing the usual academic paths in US or British universities that would have required her to focus on a more traditional and narrow form of research, she decided instead to write novels, largely based on oral histories she'd collected from her own people. Brodber's critique of the narrow focus of university history education in past generations highlights the fact that there are many stories that have been overlooked or left out. Challenging the university from outside its walls in this way can also be a form of love for the idea of it, goading the university to enlarge its vision and remain true to the values it claims to represent.

~~⟶ ⟵~~

Not far from where I live there was once an open field next to a shopping mall. I read an article in a local newspaper that described how, every few months, a truck would arrive loaded with lumber, and an old man would be there to receive it. Then, all by himself, he would set about building what I like to think of as his university. A frame went up here, a doorway there, a window over there. After a few days his wife would come. She would follow the lumber truck back to the building site, apologetic and thankful that the company was willing to take back most of the delivery. Then she would gather up her husband to go back home, where he began to dream once more of his university. Like him, we need to continue the essential work of dreaming about the kinds of universities we will want to build. Our civilization demands it.

CHAPTER 15:

A Return Volley, Forty Years Late

At its best, science fiction is not about amazing technologies and alien worlds, but about what it means to be human.

As an undergraduate, I once took a course on the 20th century American novel and had the temerity to ask my professor why we hadn't read a work of science fiction. Her reply stung: 'Science fiction is not a true literature, because it does not explore what it means to be human.'

I nodded as if I understood her point. In fact, I was stunned into silence. The rest of the class waited patiently to return to discussing the role of ennui in Walker Percy's *The Moviegoer* (1961), a book I had, in fact, enjoyed, like all the other books we'd read in that class.

Kurt Vonnegut once said erotic literature and science fiction had a lot in common, in that they are both fantasies. In one, beautiful people want to have sex with you. In the other, super-intelligent aliens want to have a conversation. Both are a kind of imagined intimacy with the Other, and it's true that some adolescent boys find the idea of interstellar war less terrifying than talking to a woman. Vonnegut went on to say that he didn't mind having his writing put in the science fiction file drawer, but what bothered him most was that some critics confused that file drawer with a urinal.

Suffice to say that science fiction has traditionally held an ambiguous position in the literary canon. Some critics, like my professor, hold the view that science fiction is an adolescent form of escape, a temporary respite from the real work of becoming an adult. At the time, I didn't know how to respond to her dismissal—though I know now that her opinion is not shared by many literature professors, some of whom I count among my close friends. Those who love good storytelling can enjoy a well-crafted space opera or time-travel yarn as much as anything else, and some even teach courses and write critical articles and books about them. Science fiction is now a recognized subfield of literary and cultural studies. Yet back in the day, science fiction was still viewed as pulp trash by many scholars.

Forty years later, I know what I should have said in reply to the put-down: 'Science fiction—at its best—is almost entirely about what it means to be human.'

Science fiction helped awaken my imagination at a critical time in life. Growing up as a working-class kid in Baltimore during the Vietnam War, it looked like I was destined for the front as I aged out of adolescence. In the end, I missed out on the draft by a few anxious years. But the thought of being marked out narrowed my vision, turning the foundations of the world to ash beneath me.

Around this time, novels like *Dune* (1965) or *The City and the Stars* (1956) enlivened me to the notion that the universe is a very, very large place, and the distant stars were places we might one day visit. It was a balm to think that we might not only build new civilizations, but new centers of human meaning. The Apollo landings on the Moon gave concreteness to my interest in science and astronomy, and eventually determined my vocation in physics.

Its advocates like to say that science fiction is a literature of ideas, perhaps even the most philosophical of all genres. There is some truth to this. While fantasy writing tries to rekindle our sense that the world just might contain magic, science fiction tends to interrogate how our penchant for invention might be the undoing of our species; as our machines gain in power and subtlety, our designs reveal more fully who we are, and what we truly desire. Other forms of literature might be more character-driven, more interested in the interior lives of people and the nuance of experience. Successful science fiction writers who talk about 'world building' as an essential part of the craft often accept that their characters might not be as richly drawn as in, say, a modernist novel by Virginia Woolf. Yet both forms of literature are interested in the fundamental texture of our shared humanity.

Even popular forms of science fiction, like the *Star Trek* franchise, explore the question of what it means to be human. For years, the television show was a target of snarky criticism by the sci-fi genre gatekeepers of the written form, who felt that the TV show wasn't really, not truly, in the purest sense, science fiction. Instead, it was a weekly installment of screenwriter Gene Roddenberry's cycle of morality plays. Roddenberry had started in Westerns, and when that genre began to lose viewer interest with the coming of the Space Age, he pitched a show that even he described as 'Wagon Train to the Stars'. Following in the footsteps of Rod Serling's *Twilight Zone,* Roddenberry found science fiction to be a genre where social criticism was safer because it could be set in outer space in some distant future. Many episodes broke taboos and dissected contemporary anxieties. There was the episode with the first bi-racial kiss during the decade of Civil Rights.

There was the one about Mutually Assured Destruction, when nuclear-tipped missiles that could cross intercontinental distances in half an hour were still new enough that we couldn't ignore them. Fans of the written form might have felt the show wasn't subversive enough for their tastes and was softened too much around the edges, but this overlooks the fact that Roddenberry had to please not only fans, but corporate sponsors and network censors.

Many other science fiction stories have entered the collective poetics of modern life. H. G. Wells' *War of the Worlds* (1897) imagines an Earth invaded by a Martian civilization that is trying to escape its own dying planet and to take over our living one. The Martians are led by 'intellects vast and cool and unsympathetic' and are bent on genocide against the human race. When the book was broadcast as a radio play on Halloween in 1938, the actor Orson Welles chose to begin the story as if the Martians had landed in a field in New Jersey, interrupting a music broadcast with breaking news of invasion. It was a brilliant reimagining of the classic novel, fun to listen to even today, and it proved so well done that some listeners panicked and believed the alien attack was real.

War of the Worlds was written at the height of the Age of Empire, and on re-reading it today it's clear that the book is a story about what it's like to be colonized, of being *in the way* of someone else's plans for the world. *The Time Machine* (1895), Wells' other masterpiece, is an imaginary vehicle for exploring the end state of a far-future society that has separated into a worker class and a leisure class. This is a society where—in a fantastic literary move—Wells turns the social hierarchy on its head: instead of the Edwardian rich metaphorically feeding on the poor, the worker Morlocks literally feed on the effete Eloi. Neither novel is an escape from human concerns, but an attempt to view them as lying within enlarged frames of reference.

And then there is the ur-story of *Frankenstein* (1818), the book by Mary Wollstonecraft Shelley that effectively launched the entire literary genre of gothic horror. When Nick Dear and Danny Boyle mounted a stage adaptation in 2011 for the National Theatre in London, they dug back into the original book as their source of inspiration—not the movies, camp and fun as they might be. In the iconic 1931 film, the monster, played by Boris Karloff, had lost his voice and was portrayed as a hulking brute. The 'monster' of the 2011 play, as in the original novel, is a fast learner, potentially capable of wonders if only he hadn't been rejected by his creator and human society more generally. As in the book, Doctor Frankenstein's creation spends much of the play chasing his creator, trying to find an answer to the question: 'You created me. Why don't you love me?'

The 'monster' here is still ungainly, ugly, feared by the one who made him. But something more is going on, in that Doctor Frankenstein has come to hate *himself* for creating the monster. This is a story about the rejection of Frankenstein's Daedalus drive to learn the secrets of the universe, to become like God in the ability to breathe life into dead matter. The Doctor ultimately comes to reject this deep desire to overcome human limitations. In a brilliant move, the director of the National Theatre production Danny Boyle had the actors playing Doctor Frankenstein and his monster (Benedict Cumberbatch and Johnny Lee Miller) switch roles from one performance to the next, to reinforce the mirror-like aspects of the characters. Who is the real monster, the creation or the creator?[1]

Humans have always had a troubled relationship with our creations, sometimes loving them too much, other times rejecting them as freaks. A truly new creation often seems uncanny because it doesn't fit within the established order of things. And when the creation itself has the potential for agency, terror flows from the possibility it might turn around and *reject us*. Shelley's story presages many of the more modern tales of AI run amok, killer robots, biotechnology gone wrong, designer pathogens let loose. Here science fiction reflects very real current human anxieties.

Going back even further than *Frankenstein*, Johannes Kepler, the astrologer, astronomer, and mathematician, is sometimes credited with writing the first science fiction novel, *Die Somnium* (The Dream) (1634). Kepler led a tragic life in many ways. His abusive father was a mercenary who beat his wife and children when he wasn't away fighting. As a child, Kepler was shy and socially awkward, and he would hide under the stairs when his father was home. Later in adult life, his mother was tried as a witch and Kepler, by that time already famous, had to defend her against a fiery death at the stake. For much of his life, the continent of Europe was wracked by brutal and deadly wars of religion, and yet the dreamy Kepler pursued a life pondering the heavens. For him, astronomy was a respite from a dangerous world. The book where he announced the third of his laws of planetary motion, the one that reveals that the square of the period of a planet's orbit is proportional to the cube of its major axis—that book was typeset in a shop embedded in the walls of a city under siege, the sound of cannonballs shaking the very ground beneath him at work. He named that volume *Harmony of the World* (1619). He was, even to the very end of his life, a dreamer.

The later *Die Somnium* was a project that Kepler variously abandoned and returned to, working on it for over forty years. It was published only after his death, and only then as a way for his family to raise money and thereby avoid poverty. The book concerns an imaginary voyage to the Moon, affording Kepler a chance to explore how the Earth might appear to someone looking down from above—far away from the tumult of war, disease, grief, and death that life on Earth had seemed to him. Was Kepler too embarrassed to publish his fantasy while still alive? Or did he take comfort in the time he spent while writing it? Perhaps both.

Science fiction, then, is not a diversion *away* from things of importance, but an orientation *towards* those very things via the vehicle of storytelling. It serves as a fodder for rumination, a stimulus to the imagination, and something to kick us out of the everyday, habitual ways of thinking that we all fall into. However, this effort to provoke the reader can sometimes take over the narrative and make it hard for other aspects of the work to shine through—the subtlety of characterization, the vividness of the setting, and any other aspects usually used to assess the quality of a piece of literature.

The novelist Amitav Ghosh has written about the inverse of the problem. In *The Great Derangement* (2016), he asks why one of the most serious problems that our civilization faces—accelerating climate change—is almost entirely absent from 'serious' realist literature. This is not a criticism of the literary community on Ghosh's part, nor a charge that they aren't paying attention to the world they live in. Instead, *Derangement* is a long rumination upon why it's so hard for a literary novelist to pull it off. He admits that he's tried himself to bring climate anomalies into his realist literary novels and found it doesn't work. Although climate change is the backdrop of our current reality, it doesn't feel believable.

Admittedly, there is now a burgeoning subgenre of science fiction—'cli-fi'—devoted to climate catastrophe. Yet Ghosh argues one aspect of climate change that makes it hard to capture in more traditional literary forms is the intrusion of the uncanny into everyday life, a sense that a previously stable reality is shifting beneath and around us. Familiar things turn strange. The rivers of Europe and the US West shrink, uncovering shipwrecks and dead bodies long forgotten. The fires of Northern California make the sunset skies of San Francisco look like Mars. A pandemic shuts down the world economy and wild animals return to city centers. The Pacific Northwest sees temperatures like those of the Saudi deserts. The catastrophic nature of climate change, when it plays a role in driving plot or character motivation, can easily swamp the smaller parts of the story and overtake other narrative devices, turning a realist literary novel into fantastical literature.

Philp K. Dick's *Do Androids Dream of Electric Sheep?* (1968) is about a world where urban-dwelling humans live in a ruined environment and yearn to reconnect with the living world. Here, machines have become almost indistinguishable from living things. But the biosphere itself is completely impoverished, so they must settle for ersatz bionic replacements—and one's place in the social hierarchy is determined partly by the size of the robot you can afford. If you can only afford a robotic rat, you're looked down upon by your neighbor who has a robotic sheep. The boundary between living and nonliving things is blurred, to the extent that the androids (i.e., *humanoid* robots) don't always know they are androids. The Ridley Scott movie adaptation of the novel, *Blade Runner,* is in many ways less subversive than Dick's written story. But even there, in the guise of a near-future thriller set in the year 2019 as imagined in 1982, *Blade Runner* confronts our fear of the machine Other, and the social crisis that would be brought on by the creation of superior androids that can pass as one of us.

The New Wave of the 1960s, and the subsequent parade of punks, cyberpunk, steampunk and now solarpunk, as well as what came to be called 'slipstream' authors such as Samuel R. Delany, Octavia Butler, Ursula K. Le Guin, James Tiptree (Alice Sheldon)—these writers reinvented the old science fiction tropes in order to nudge readers to think about race, gender, neurotype and sexuality. Later writers such as William Gibson, Kim Stanley Robinson, Nnedi Okorafor and N. K. Jemisin expanded on these themes to include religion, and culture, along with all the various forms of oppression and slavery that are based upon an alienness in our hearts that is all too human. All these writers were poking and prodding at the question of what separates us from one another—pulling at the scar forming around the wounds of modern life, among them the itch for chaos.

As the space opera and techno-thriller films have resurged in popularity in the last few decades—think of *Babylon 5,* the *Battlestar Galactica* reboot, *Moon,* the *Expanse* and *Sense8* series—they have taken a far more multifaceted approach to story, refusing to take much for granted, because readers and viewers will no longer allow it. This is because, at its heart, science fiction has always been about the diversity and complexity of the human condition. It just approaches that subject by forcing the frame farther and farther outward until it almost breaks, by embedding that question into the long arc of deep time, by placing humans in the context of the star stuff of which we're made, and by highlighting all our entangled forms of co-evolution with the living and nonliving world—with animals and humans, with machines and all our other inventions, including societies and cultures. Science fiction takes all of these as its working medium, its playground of ideas, and all that serious play is in service of the question: what does it mean to be human in such a world?

That's what I should have said to my professor, all those years ago. I'm sorry I'll never get to tell her.

CHAPTER 16:

All Our Plans Will Go Awry

Accepting the failures of the past can emancipate our imaginations. The future is dark, but ready to be lit up by new discoveries.

❦

'Our business in this world is not to succeed, but to
continue to fail, in good spirits.'
— **ROBERT LOUIS STEVENSON**[1]

In April of 1915, Virginia Woolf wrote in her journal: 'The future is dark, which is the best thing the future can be, I think.' Perhaps it's best that we can't see what's coming. The lack of certainty gives us some freedom to act. Or perhaps it's truer to say that it lets us believe we have such freedom. A stubborn insistence on our role as active agents in the world is a good thing, surely—even when we so often seem to bungle things, despite our best intentions.[2]

We might think of our lives as a tree. The future lies above us, the past down below. Looking up, we can see all our manifold futures branching outwards, in ways that are contingent upon unpredictable events: natural disasters, emerging epidemics, new discoveries and inventions, revolutions and societal collapse. In the canopy far overhead, at the end of any single branch lies a leaf, one of hundreds of thousands. Each leaf is a possible future, alive, growing, spreading as we watch.

We stand beneath this superabundant tree of possible futures, in the dappled shadows of the present. It overwhelms the imagination, and the future looks dark. What to do?

The past also confounds any naive desire for it to contain a single compelling story. When we cast our thoughts beneath our feet, we find rooted histories that feed the present. But the closer we look, the more we see networks of cross-connections and multiplying perspectives. There are always more narratives to find, and new ways to tell those histories. While the past might be at rest, our understanding of it shifts as we excavate its meaning.

A Chinese proverb holds that if you live long enough, you will get to see the bodies of your enemies float down the river. In the same sense, although George Orwell believed that history was written by the winners, it's also true that those who lose for a time can prevail in the long run. I once looked at a 1950s-era course catalog for my university, William & Mary, which revealed that all the history courses were about either American or British history. That's no longer the case: my colleagues now cover, among other areas, the social history of early modern Spain, women's and gender history, the early stages of the Scientific Revolution, labor history and the history of financial panics, as well as regional histories spanning from Latin America to Africa. Across campus, historians can also be found in departments like philosophy, music, art and economics. None of these scholars would consider there to be one single 'History' with a capital H.

My colleague, the historian Nick Popper, once told me that he hopes one day to teach a course he wants to call the 'History of History Courses'. From him, I learned that the notion that all

American college students should take an introductory course about American History first came into vogue around the time of the First World War, where those courses were intended to teach prospective soldiers what they were fighting for. These days, history departments would likely rebel if called upon to teach courses with the aim of inculcating nationalism in their students. While many young people arrive at university believing that knowledge of the past is fixed in amber, they are challenged to see that history is still being written, to always consider who is telling the story, and to keep an open mind. During one heated faculty discussion we had about proposed curricular reforms, a physics colleague of mine, worried about the proliferation of historical viewpoints and with an eye to preserving 'the canon', opined that history should be about 'finding out what really happened'. The professional historians in the room chuckled the loudest.

Like the future, the past is shrouded in shadows too. But the realization that our understanding of the past is fluid can emancipate our imaginations, as it frees us from the tyranny of believing there will only be one kind of future.[3]

~~~~~

When I bought my home in Virginia in 1985, the deed had a covenant placed on it by the original owner in the 1930s, barring the sale of the property to anyone of 'African or Hebrew descent'. Through such legal means, dead men and women could limit the life prospects of people who lived long after they passed from the scene.[4]

While the idea that we're creating a better world for our children is a common refrain, the fear of personal extinction can sometimes overshadow our more generous impulses. It might seem paradoxical, but arguably it's only by internalizing our own inevitable passing from the scene that we can become more objective about how to help future generations.

The acute crisis of the Covid pandemic demanded that we focus all our energies upon it. The climate crisis has a different character, a slow and smoldering burn compared with the pandemic's raging wildfire. But the two crises are similar in that we cannot really 'solve' them. It's in their nature that they are something we must learn to navigate, adapt to, and mitigate. We must learn from our past mistakes even as we inevitably make new ones.

Some of what we learn from the pandemic can be transferred to the climate crisis. For example, the need for more resilient designs for our critical networks such as food, water, energy, information, and supply chains. The need, too, to do a better job of caring for the most vulnerable among us. But the global and enduring character of the climate crisis makes it especially hard to see what might be coming by the end of the century. Climate change requires us to imagine possible futures in which the world becomes unrecognizable. This is not only because of changes in the natural environment, but also because of transformations in human values and aesthetic judgements, all of which will drive social and political reformations.

Coming to terms with one's own death is hard work. What about the death of certain collective futures? That requires, in part, a grieving for all those imagined possibilities which reveal themselves to be false dreams. The dream of the socialist worker's utopia, or the fantasy of a frictionless capitalist marketplace. The libertarian minimalist state, or the maximal nanny-state that cocoons us in its warm embrace. Those dreams are too simple, too rigid, too universalized to be humane. Old ideologies are based upon outdated intuitions about what makes people tick, the nature of the material world, and our place in it. In the end, none of them in their 'ideal' form is nourishing and robust enough for the long haul. Instead, nature always favors mixed and hybridized strategies for ecosystems, nimble and open to change. This seems true for political economies too.

Acceptance that the fossil-fueled world is coming to an end doesn't mean that human civilization is heading into terminal decline. Along a more hopeful road, in the coming decades and centuries, there will still be families and learning communities, politics, arguments and reconciliations. There will still be cities and commerce, and love, music, friendship, and art. There will be complex and vibrant civilizations, restless and muddling through. If a time machine dropped us down in their midst, we'd probably feel out of place at first, missing some of the familiar markers of everyday life. But there is no reason to believe that our current limited menu of encrusted ideologies will order the geopolitics of the future world. At least, let's hope they won't, because they're certainly no longer fit for purpose.

We shouldn't get too downcast that we haven't figured it all out yet. We should strive to be happy warriors in all things, and to see that we can learn from our failures. For all its plaudits, my own field of physics has failed in big ways. We've failed to unify the forces of nature, failed to explain how quantum mechanics can be made consistent with general relativity. Failed to explain dark matter and dark energy. And a little closer to home for me, the nuclear fusion program has so far failed, even after generations of effort.[5] But scientists don't view these failures as a reason to give up. Instead, they are an invitation to the next generation, a sign that great discoveries and achievements could be in the offing. Properly recognized as a goad to new effort, those failures can even be energizing. In fact, prospects for fusion energy appear brighter now than ever before, and serious private money is flowing into a variety of startup companies.[6] The future is dark, but ready to be lit up by new discoveries. We just don't know when or how.

We can and must let go of the illusion of control, trusting instead to the human capacity for kindness, invention, and the shared creation of meaning. Rather than building upon old and outdated social and political foundations, a future could be entwined with a wilder view of the world and our place in it, a view that loves the ingenuity of both nature and human invention. We will love them all because they will have become beautiful and worthy of love. We can hope that will be the case, anyway.

How to improve our chances of realizing this world? When, how, and what permanent changes the pandemic will leave in its wake, we are still guessing at. Our children and grandchildren will know far better than us how the current social, economic, technological, and climate trends will play out. So, our wisest course in the here-and-now is to make choices that will give future generations as many options as possible. They must be able to overwrite our current plans for them, if need be. That will only be possible if our plans of today allow for their agency tomorrow.

❦

George Orwell opened his essay 'England, My England' (1941) with the sentence: 'As I write, highly civilized human beings are flying overhead, trying to kill me.' It reminds me of an incident that occurred when I was a young researcher, in the mid-1980s, attending my first international workshop in the south of Italy. It had been an amazing week, meeting many of the most famous people in my field, but I also got to meet another young mathematician. He was from Viet Nam. This would have been only a decade or so after the US had ceased hostilities there, but long before the normalization of relations.

We were still in the grip of the Cold War, and the young mathematician had come by way of Moscow. The two of us would chat over coffee or sit with one another in the evening to share a drink, to talk of our mutual interest in mathematics and physics. As the week wore on, he finally began to relax a bit, and told stories of growing up in Hanoi during the bombings. Night after night, on again and off again, over some years, the US sent B-52s in a vain attempt to force North Viet Nam to give way at the negotiating table. Growing up in the US, I remembered hearing of those bombing raids, and the nightly reports on the news about the war. But here was a young man of a similar age to me, who'd lived in a bunker, his mother cooking rice over a small flame, while highly civilized men, my countrymen, had been trying to kill him.

But we are always free to decide for ourselves what things will mean, and that freedom calls out to us especially when we are young. We must honor this capacity to decide on new directions as our most important bequest to the next generation. On the last night of the conference, as the sun was sinking on the horizon, at our closing reception, he gave me a gift, a bottle of Russian vodka that he'd carried in his suitcase to give to an American friend.

Virginia Woolf's insight still stands: the future is dark not because it is bleak, but because it is unknowable. We need to make it easier for future generations to do the rewriting that will be necessary. They need the freedom to make new friends, build new cities, to found new universities, to walk the dunes of Mars, to re-wild the rivers of Earth that past generations tried so hard to tame. Any plans we make will go awry; what our children need more than anything else is room to maneuver.

CHAPTER 17:

A Science Without Time

The flow of time is central to human experience.
Why isn't it central to our physics?

❧

'What then is time? Provided that no one asks me, I know. If I want
to explain it to an inquirer, I do not know.'
—SAINT AUGUSTINE[1]

'Out of fear of dying the art of storytelling was born.'
— EDUARDO GALEANO, *MIRRORS: STORIES OF*
ALMOST EVERYONE (2009)

I have a memory, a vivid one, of watching my elderly grandfather
wave goodbye to me from the steps of a hospital. This is almost
certainly the memory of a dream. In my parent's photo album of
the time, we have snapshots of the extended family—aunts, un-
cles, and cousins who had all traveled to our upstate New York
farm to celebrate my grandparents' 50th wedding anniversary. I
am in some of the photos along with my brother, a pair of small

faces mingled with smiling giants. I remember the excitement of the evening, being sent off to bed but then staying up late at the top of the stairs, listening to the pleasant babble of adult voices. I have no recollection of what happened later, but it did not involve a timely visit by me to the hospital. My father told me many years afterward that my grandfather took ill that night and was rushed to the emergency room, where he died on the operating table.

The memory of my grandfather's farewell still provokes in me a longing for a world where a more lawful order holds, where connections with those we love are not bound by time and space. A central purpose of early science and philosophy was to satisfy such longings: to get off the wheel of time and life to which we are bound and to glimpse what the French-born writer George Steiner has called a 'neighboring eternity'. But our human sense of time is that we are bound by it, carried along by a flow from past to future that we cannot stop or slow.

The flow of time is certainly one of the most immediate aspects of our waking experience. It is essential to how we see ourselves and to how we think we should live our lives. Our memories help fix who we are; other thoughts reach forward to what we might become. Surely our modern scientific sense of time, as it grows ever more sophisticated, should provide meaningful insights here.

Yet today's physicists rarely debate what time is and why we experience it the way we do, remembering the past but never the future. Instead, researchers build ever-more accurate clocks. As I write, the current record-holders reside in the laboratories of Professors Jun Ye at the University of Colorado and Hidetoshi Katori at the University of Tokyo, who shared the three-million-dollar 2022 Breakthrough Prize in Fundamental Physics for their work. Their clocks measure the vibration of atoms that form what's called an 'optical lattice'. These devices are accurate to 1 second in 15 billion years, roughly the entire age of the known universe.[2] Impressive, but it does not answer 'What is time?'

To declare that question beyond the pale of physics doesn't make it meaningless. The flow of time could still be real as part of our internal experience, just real in a different way from a proton or a galaxy. Is our experience of time's flow akin to watching a live play, where things occur in the moment but not before or after, a flickering in and out of existence around the 'now'? Or is it like watching a movie, where all eternity is already in the can, and we are watching a discrete sequence of static images, fooled by our limited perceptual apparatus into thinking the action flows smoothly?

The Newtonian and Einsteinian world theories offer little guidance. They are both eternalized 'block' universes, in which time is a dimension not unlike space, so everything exists all at once. Einstein's equations allow different observers to disagree about the duration of time intervals, but the spacetime continuum itself, so beloved of *Star Trek's* Mr. Spock, is an invariant stage upon which the drama of the world takes place. In quantum mechanics, as in Newton's mechanics and Einstein's relativistic theories, the laws of physics that govern the microscopic world look the same going forward or backward in time. Even the innovative speculations of theorists such as Sean Carroll at Caltech in Pasadena—who conceives of time as an emergent phenomenon that arises out of a more primordial, timeless state—concern themselves more with what time does than what it feels like.[3] Time's flow appears almost nowhere in current theories of physics.

For most of the past few centuries, conscious awareness has been considered a problem too hard for physics to tackle, postponed while we dealt with other urgent business. As scientists drove ever deeper into the nucleus and out to the stars, the conscious mind itself, and the glaring contrast between our experience of time's flow and our eternalized mathematical theories, was left hanging. How did that come to pass? Isn't science supposed to test itself against the ground of experience? This disconnect might help to explain why so many students not only don't 'get' physics but are positively repulsed by it. Where are they in the world picture drawn by physicists? Where is life, and where is death? Where is the flow of time?

The Atomist Greek philosopher Democritus had already pointed out the conundrum here, back in the 4th century BCE. By careful observation and reasoning, he argued, we come to the conclusion that the senses can fool us, but it is through evidence from the senses that we have come to that conclusion. This realization leads to a sophisticated philosophical understanding of how we come to know things about the world: not by trusting our senses naively, but by testing our thoughts about how the world works empirically. It is an insight that has borne tremendous fruit, yet one that counsels perennial humility.

The phenomenology of experience, such as our internal perception of the passage of time, is an area owned by cognitive science and philosophy. The exterior world is traditionally the playground of physics. Yet to separate the inner and outer realms in this naive way is misleading. It is our brain that does physics, after all. In the end, the two sides strive to find bridges between them, if only through metaphor, to find connections between the myriad ways in which humans experience themselves in the world.

One useful connective metaphor is to think of the brain as a storytelling engine.[4] In *Physics and Philosophy: The Revolution in Modern Science* (1958), the German physicist Werner Heisenberg reflects upon the fact that language and our sense of the world are interwoven. Our sensory organs and the brain are products of long millennia of evolution; our DNA is a kind of memory carried down through eons of deep time by direct lineage, parent to child, all living and dying on the same planet, all learning to survive within a narrow range of space, time and energy scales. Our genes, our personal memories and the very structure of our languages—these are all encoded forms of knowledge about the world. But this knowledge is based upon an extremely restricted range of physical experience.

Language is infinitely variable, and comes in a wide variety of forms, from the metaphorical, evocative, dreamlike and magical, to the logical, direct and tightly organized. What form of language is most useful for talking about the world beyond our everyday experience? What language can take us into the heart of the atom and beyond the edge of the galaxy, and describe the passage of time that pulls the world inexorably forward on these scales? Heisenberg argued it is the logical and mathematical language used in modern physics, precisely because that language is so rigid and formalized. When building a bridge into the dark, build using careful, sure steps. But we want to understand our own place in the world, not just how the world is out there; we also want to understand how we come to experience the world as we do. That calls for the more fluid and evocative language of poetry and storytelling.

Current cognitive science suggests that our memories are a kind of story that our brain creates, formed from the clay of sensory input, sorted into patterns based upon our past life experience, guided by predilections we have inherited in our DNA. Some of the intuitions that infuse those memories are basic to our sense of the world: the smooth geometry of three spatial dimensions; the clear and obvious distinction between before and after; the flow of time. Physics calls into question the smoothness of space and time, the psychological flow of time, and even asks: why do we remember the past but not the future?

Consider our experience of 'now'. This seems at first to be a simple thing, a well-defined point in time. We certainly seem to anticipate a particular 'now' coming at us from the future, and then receding from us into the past. Our experience of the 'now' is built out of a mix of recent memories and our current sense perceptions, what we see, hear, feel, taste and smell. Those sensory perceptions are not instantaneous, but signals from stimulated nerve endings. Those signals are sent to the brain, a dynamic network that itself has no global clock. The brain is like *The Palace of Dreams* in Ismail Kadare's 1981 novel of the same name: a massive bureaucracy, full of intrigues, gathering intelligence from the restive provinces about the Sultan's dreaming subjects in hopes of divining their intent.[5] 'Now' is a construct of this angst-ridden Sultan-brain—a local theory of what's happening, cobbled together using bits of news from the sensory hinterlands.

We usually don't sense this mingling of near past and near future because our brain works quickly enough to obscure the process. Yet there are moments when it struggles to keep up. This is why baseball pitchers can throw exploding fastballs, where the ball seems to suddenly leap across the space between the pitcher and the batter, and why batters can hit frozen rope, where the ball seems to stretch out into a line: if the ball moves too rapidly for the brain to track, the brain makes up a different story about the motion.

Sitting in the right field bleachers at Camden Yards in Baltimore, more than 400 feet from home plate, the late Kirby Puckett of the Minnesota Twins once hit a line drive right at me. Before I could move, before I could consciously perceive the ball crossing the intervening space, it was in front of me, dropping into the crowd a few rows away. As a physicist, I know the ball followed a smooth parabolic trajectory from start to finish. But that day, it seemed to leap from the bat, arriving before the sound of the hit, arriving before even the thought of it in my mind could catch up.

It's possible that our experience of the flow of time is like our experience of color. A physicist would say that color does not exist as an inherent property of the world. Rather, light has a variety of wavelengths, and things in the world absorb and emit and scatter photons, granules of light, of various wavelengths. It is only when our eyes intersect a tiny part of that sea of radiation, and our brain gets to work on it, that 'color' emerges. It is an internal experience, a naming process, an activity of our brain trying to puzzle things out.

So, the flow of time might be a story our brain creates, trying to make sense of the chaos. In a 2013 paper, the physicists Leonard Mlodinow of Caltech and Todd Brun of the University of Southern California went even further to argue that *any* physical thing that has the characteristics of a memory will tend to align itself with the thermodynamic arrow of time, which in turn is defined by the behavior of extremely large numbers of particles.[6] According to this theory, it is no puzzle that we remember the past but not the future, even though the microscopic laws of nature are the same going forward or backward in time, and the past and future both exist. The nature of large systems is to progress toward increasing entropy—a measure of disorder, commonly experienced as the tendency of hot or cold objects to come into equilibrium with their surroundings. Increasing entropy means that a memory of the past is dynamically stable, whereas a memory of the future is unstable.

In this interpretation, we are unable to see the future not because it is impossible, but because it is as improbable as seeing a broken window heal itself, or as a tepid cup of tea taking energy from the atoms of the surrounding room and spontaneously beginning to boil. It is statistically extremely, extremely unlikely.

We might also think of the entity we call 'the self' as emerging from chaos—a kind of story told with light and matter. The massive sculpture *Quantum Cloud* (1999-2009) by the British sculptor Antony Gormley stands on a pier next to the Millennium Dome in London. It consists of a dense, three-dimensional pattern of steel rods, arranged in a semi-random pattern, that appear to converge on a central, ghostly human figure. It is natural for the viewer to identify with that figure, but the self that one sees in Gormley's sculpture varies from different perspectives, so who is the 'I' in the cloud?

Similarly, our sense of self derives substantially from our memories, which seem continuous and durable. Yet those memories must emerge as a story that the brain develops from something less structured, a chaos where before, now and after have no rigid moorings.

The use of the term 'quantum' in Gormley's artwork points to physics. As the American physicist Richard Feynman noted in his 1942 PhD thesis,[7] the time evolution rules of quantum mechanics can be reinterpreted to say that particles such as electrons travel along all possible paths from their beginning to ending points, with the quantum transition rules emerging through a kind of averaging over that microscopic chaos. In this view, the world has a profligate richness of histories, each eternally present even if not perceptible to us. Feynman's 'sum over histories' interpretation is now a standard tool in fundamental physical theory and is even used in fields far removed from theoretical physics. A recent Google search on the related term 'path integral' returned almost half a million hits.

If the Feynman approach gives good experimental results—and it does—does this imply that all those histories truly exist? Most physicists believe they are struggling to understand the Universe as it is, not simply developing computational tricks that reveal nothing beyond our own cleverness. Yet few of them regard every possible quantum path to be its own, genuine reality. Somehow, only certain potentialities become realities, and large systems such as human observers are somehow swept along from past to future.

So why do we remember the past but not the future? Perhaps the answer lies in the very unpredictability and inconstancy of reality at the smallest scale. In the mind's eye of the modern physicist, even the vacuum seethes. Having spent my professional life peering at nature through the lens of theoretical physics, I no longer recoil from the thought that nature might be chaotic at heart. It seems to me now that it opens within us a glimmer of freedom, not by equating mere randomness with ersatz free will, but by reminding us that the question of our own freedom is not yet settled. In *How Physics Makes Us Free* (2016), philosopher Jenann Ismael argues that our internal sense that we have free will is fully consistent with the laws of physics, provided we think carefully about what physics is telling us, and what we mean by free will.

Lee Smolin at the Perimeter Institute in Canada argues that scientists must change tack, accepting the flow of time as real and building the church of a new physics upon that rock. The British physicist Julian Barbour takes an opposite stance: going beyond Newton and Einstein, in *The End of Time* (1999) he proposes that time itself is an illusion.[8] Instead, the Universe according to Barbour consists of a collection of static moments, like a pile of unsorted photographs tossed into a shoebox. Each photo contains a snapshot of the world entire, a unique configuration of all things: planets, galaxies, bumblebees, people. Barbour gives the collection of all possible moments an evocative name: 'The Heap'.

Because each instant in The Heap is a moment of the entire world, it contains references to many other moments, so the shoebox also contains an implied web of connections, branching threads of mutual association. Following a single thread, one would experience an apparent flow of time. Most threads would follow isolated paths that are without sense or meaning, but a very few threads and their neighbors pursue paths that are mutually coherent. We might say that such paths tell a story, or that they include a sensible memory of the past at each step. The family of threads that are mutually coherent is robust, whereas the isolated and incoherent threads are fragile, with brittle associations providing no neighboring reinforcement.

This idea is interpreted in two time-lapse videos created in a collaboration between me and Kit Tracy, my child.[9] Both videos were shot by us over a minute just before midnight on 3 June 2015 on the Las Vegas Strip; the videos contain the same set of a few thousand images. In the first, Version A, the time ordering has been scrambled; the second, Version B, gives the images in the original time order. Version A is hard to follow, because there is nothing to hang on to. Attentive viewing of Version B, in contrast, reveals a multithreaded story from the 35-second mark to the 45-second. Someone is surrounded by police. An ambulance arrives on the scene. We make a U-turn, and another policeman arrives on the 45-second mark. Given the date, the location and the approximate time, it would be straightforward, in principle, to find out more details from police reports. The time-ordered images in Version B allow us to follow a coherent narrative thread through the collection of images—Barbour's Heap—and to make deeper inferences about their meaning. There is a narrative coherence in Version B that's absent from Version A.

Barbour's Heap reminds me of a photo I found among my parent's things when they were old. Luckily my mother was still alive to talk about it at the time. In the photo, my parents are both in their 20s, not yet married, at a nightclub in New York City. It is some time in the Second World War, and they are with friends in uniform, all merchant seamen like my father. That night, he was dressed like a gigolo, clearly off-duty and having fun. My mother is shockingly young, smoking a cigarette, looking like an ingénue. The other men around the table are about to ship out for convoy duty in the North Atlantic. Within a week, all those men would be dead.

For me, that photo carries layers of memory and meaning. There is the grief my parents must have felt for their friends, people I never met, carried away by time's flow long before I was born. The direct experience of mass death during the war helps to explain the silences I experienced in our home growing up, silences I now know were full of memories. Looking at the photo today, I know who my mother and father would become, how my father would leave the merchant service and buy a farm. A city boy, gone to sea, then settled far away from the water. In that photo I think I see a glimpse of why: the memory of friends lost in the North Atlantic, drowning. I can understand my father's anxiety, his need to stand on firm ground in a dangerous world. All his life he told us stories so that he would not be forgotten.

He acted out the same eternalizing impulse as the ancient Pythagoreans, just differently expressed. He told stories out of fear of dying, like Galeano says. We tend to fence off science from other areas, imagining that a quantum wave function or a set of relativistic field equations express a fundamentally different aspect of time than the kind of time that is embodied in old family tales. In the process, we lose sight of the fact that scientific theorizing and storytelling are both, at heart, driven by a fear of dying and an itch for the eternal.

Before I am carried away by time's flow, I want to share one last memory, again as a small child. I am sitting in a warm sunbeam on the living room floor of our farmhouse, watching the gentle chaos of drifting dust motes, small worlds entire, next to my sleeping dog, King. We were—are—will be—best friends forever. Always at peace.

Epilogue

The Marshall Islanders like to tell stories about their ancient prowess as warriors. Upon the arrival of Europeans in the 16th century, if any ships ran aground the crew would be killed and the cargo taken. But in 1883 the Rainier, an American merchant vessel bound for Japan, struck the reef at Ujae Atoll, breaching the hull and forcing the sailors to abandon ship. The daughter of the captain, newly married to the first mate, was bound with ropes to a large armchair by the crew, hoisted over the side, and planted with much fanfare into a lifeboat adrift in the heaving sea. The sailors offered to blindfold her, fearing they were about to be overrun by brutal savages. But the woman said she was not afraid and insisted that she wanted to see what the islanders were going to do to her. She was rowed to shore wrapped in an American flag.[1]

In the end, all aboard the Rainier were given shelter. As the Ujae Islanders tell it now, in stories and songs carried down through the generations and celebrated in an annual festival, the wreck of the Rainier has become like a fairy tale.[2] The Ujae ruler was a kind man, and the Americans were all taken in and looked after—no small sacrifice given the limited food and fresh water available to the several hundred islanders. In their telling of the story, they welcomed those in danger, refugees, who in turn shared their provisions with them. The islanders took care of the castaways over many months until a rescue ship arrived.

The tale ends with a modern coda. During the Second World War an American naval vessel arrived at Ujae. A landing craft from the main ship rode up the beach and an officer stepped off. A descendant of a survivor of the wreck, the naval officer wanted to thank the people of Ujae for rescuing his ancestor, without whom he would never have been born. He was closing the loop of human kindness across the generations, keeping faith with a parallel narrative tradition, passed down from parent to child on the far side of the planet.

And then the crew and the islanders celebrated with a party and played baseball.

ACKNOWLEDGMENTS

These essays were composed over a number of years, more than I'd care to count. That means there are many people who I need to thank, either for ideas, inspiration, encouragement, or just for putting up with my shifting interests and too-short attention span. As a faculty member, I was blessed with the opportunity to work with talented and dedicated colleagues, to form lifelong friendships, and to be reminded that undergraduates get younger every year. I'd particularly like to call out my friends Bill Cooke and Dennis Manos who put up with my running commentary on whatever it was I was interested in at the moment and would always be ready with a wry comment and good humor. They helped to remind me that, although I might sometimes leap the corral of academia and go free range, I am at heart and by temperament a scientist.

Writing can be a solitary avocation, but friends who are willing to read drafts, suggest improvements, or just to put up with you when your mind might be elsewhere, are golden. Feedback on the essays when I first tried my hand at writing for a general audience were extraordinarily valuable. Special thanks to my friend Carey Bagdassarian who, like me, is a physicist who also loves writing and reading, and a companion in the adventure of self-publishing. You got there first, Carey, and it looks like the waters aren't so cold after all. That gives me courage. Teresa Longo was also part of my first-ever writers group. She was a kind and perceptive reader who asked probing questions in a way that always made me believe there was so much more to say.

Colleague Barbara King indulged me in a very memorable, and very long, conversation about AI ethics and human intimacy which set some of the ideas in motion that led to the essay

'The Yowl! In the Machine'. Deborah Morse first alerted me to the emerging field of animal studies, and Kelly Joyce did the same by flagging for me the importance of asking how algorithms reflect our ethical values. Robin Cantor-Cooke, Arthur Knight, Steve Otto, Martha Howard, and Leisa Meyer read early drafts of some of my writing and Tom Peyser who read them all. Thanks to you all for your feedback and encouragement, and most importantly for your friendship.

Corey Powell and Sally Davies were my editors at *Aeon*, and they helped me become a better writer and find my voice. Sally is amazing, and I was very lucky she was available and willing to help me turn these essays into a book. Thank you, Sally!

Ross Andersen at *The Atlantic* gave early encouragement to keep writing, along with Meghan Houser at Crown. Meghan: thanks also for connecting me with Grace Ross at Regal Hoffman & Associates. Although we parted ways in the end, it was fun while it lasted.

The William & Mary Center for the Liberal Arts funded my trips to the Marshall Islands in 2015 and the Mississippi Gulf Coast in 2016, which shaped so much of my thinking and later writing. In the Marshall Islands I'd particularly like to thank Alson Kelen, Kathy Jetnil-Kijiner, Jack Niedenthal, Dr Kondrad Hayashi, Desmond Doulatrum, Makarusa Porosotano, Melvin Mellan, and Rosety Keju. While they are not mentioned by name in these essays, our conversations nudged me in the right direction at a time when I was struggling to find the way forward in my writing. Sometimes it's good to find your worldview unsettled, and these people helped me to better understand how intertwined climate change and indigenous rights are, and how vulnerable communities call us to follow the better angels of our nature. I would especially like to thank Robert Chiu and Dr Seyoume Teshome, who took me under their wings during my visit to Majuro and introduced me to many of the others I've mentioned above. I'm grateful to all of them for their kindness to a stranger.

The folks at the Santa Fe Science Writing Workshop, especially George Johnson and Sandra Blakeslee, provided important feedback on some earlier material and essay ideas. Lon Otto at the Iowa Summer Writing Festival also gave insightful feedback, encouragement, and concrete suggestions on how to become a better writer.

Thanks to Doug Workman who first suggested that I try self-publishing, and for Maciek Sasinowski who showed me how.

And, finally, I'd like to thank my wife, Maureen, whose support, encouragement, and love have carried me through more than four decades of our life together, a form of undeserved grace from the universe. And, my child, Kit, who is an ongoing inspiration and goad, reminding me that while I might be growing old, the world is forever young.

PUBLICATION HISTORY

"Sky Readers" first appeared in *Aeon* in December 2015, edited by Corey Powell.

"A Science Without Time" first appeared in *Aeon* in April 2016, edited by Corey Powell.

"Learning to See" first appeared in *Aeon* under the title "Behold: Science as Seeing" in May 2018, edited by Sally Davies, and it was reprinted in slightly modified form under the current title in the July-August 2018 issue of *American Scientist*.

"How Much Can We Afford to Forget, if We Train Our Machines to Remember?" This essay first appeared in *Aeon* in April 2019, edited by Sally Davies. It was reprinted in the Winter 2020 edition of *Lapham's Quarterly*.

"A Return Volley" first appeared in my blog *The Icarus Question* in March 2019 and was republished by *Scroll.in*, where it appeared under the title "Don't listen to the critics—science fiction explores what it means to be human in the truest way."

The Introduction and the essays "The Mind Shaper and Maze Maker," "Hooked on a Feeling," and "All our Plans Will Go Awry" appear here for the first time. Unless otherwise noted, the remaining essays first appeared in my blog and have been completely revised and updated for this collection.

Any errors or omissions in these essays are due to the author alone.

ABOUT THE AUTHOR AND EDITOR

Gene Tracy is Chancellor Professor of Physics Emeritus at William & Mary in Virginia, where he was the founding director of the Center for the Liberal Arts. His essays have appeared in places including *Aeon*, *Lapham's Quarterly* and *American Scientist*. He blogs at *The Icarus Question* (https://gene-tracy.com).

Sally Davies (editor) is a writer and editor whose interests include science, philosophy, feminism and the arts. She is Senior Editor at *Aeon* and was formerly digital editor of *FT Weekend* as well as the technology and innovation correspondent for the *Financial Times*.

NOTES

Preface

1. Strevens, Michael. *The Knowledge Machine: How Irrationality Created Modern Science.* Liveright, 2020.

1. The Mind Shaper and the Maze Maker

1. The stories the Marshallese tell of these events can be found in Tobin, Jack A. *Stories From the Marshall Islands: Bwebwenato Jan Aelon Kein.* Amsterdam, Netherlands, Amsterdam UP, 2001. In particular, see "5. Origin of the *Irooj* (Chiefs) of the Marshall Islands," p. 51; "6. Liwātuonmour," p. 53; and "The story of Liwātuonmour and Lidepdepju," p. 54.

2. Tobin, Jack A. *Stories From the Marshall Islands: Bwebwenato Jan Aelon Kein.* Amsterdam, Netherlands, Amsterdam UP, 2001. In particular, see "8. About a woman named Lōktanūr," p. 56.

3. Sky Readers

1. Jeffers, Robinson. *Roan Stallion: Tamar and Other Poems.* Rogue Scholar Press, 2022.

2. Seeker. "How GPS works." *YouTube,* 21 Oct. 2013, https://www.youtube.com/watch?v=IoRQiNFzT0k (Accessed November 12, 2015).

3. Gibbons-Neff, Thomas. "After Two Decades, Sailing by the Stars is Back at the Naval Academy," *The Washington Post,* 13 Oct. 2015.

4. See, for example: Kandel, Eric R. *In Search of Memory: The*

Emergence of a New Science of Mind. WW Norton, 2007. The connection with emotional charge and memory formation starts on page one. The connection of memory to forming spatial maps is discussed in the chapter 'A return to complex memory'.

5. Finney, Ben R., and Eric M. Jones, eds. *Interstellar Migration and the Human Experience.* Vol. 4. University of California Press, 1986. This is a collection of articles from a workshop held at Los Alamos National Laboratory.

6. In addition to the book *Interstellar Migration and the Human Experience,* already cited, the writings of Ben Finney are a rich source of information about Polynesian voyaging. I have drawn most heavily here upon *Voyage of Rediscovery: A Cultural Odyssey through Polynesia* (1994), and *Pacific Navigation and Voyaging* (1976). A good general source on the Pacific Island cultures is Kirch, Patrick V. *On the Road of the Winds: An Archaeological History of the Pacific Islands Before European Contact.* University of California Press, 2017. I want to thank my colleague Prof. Jennifer Kahn for these suggestions.

7. See, for example, Riesenberg, Saul H. "The Organization of Navigational Knowledge on Puluwat." *The Journal of the Polynesian Society,* 81.1, 1972, 19- 56. In other parts of the Pacific, like the Marshall Islands, navigators also use something called 'wave piloting'. This technique exploits the regularity of ocean swell between the islands, and a skilled navigator can tell the direction of land just by the motion of the boat. See, for example, Tingley, Kim. "The Secrets of the Wave Pilots." *New York Times Magazine,* March 17, 2016.

8. The material about the great observatories in the 19th century is drawn largely from Schaffer, Simon. "Astronomers Mark Time: Discipline and the Personal Equation." *Science in Context* 2.1, 1988, 115-145.

9. Clark, Andy, and David Chalmers. "The Extended Mind."

Analysis 58.1, 1998, 7-19.

10. The quote from Morrison can be found in Cousins, Norman, et al. *Why Man Explores: A Symposium Held at Beckman Auditorium California Institute of Technology Pasadena California July 2, 1976.* Washington, National Aeronautics and Space Administration, 1977. Series EP123. SuDoc number NAS 1.19:123. Available online through the NASA History Office https://permanent.fdlp.gov/lps70079/lps70079/history.nasa.gov/EP-125/ep125.htm.

11. Wikipedia contributors. "Geography of Pluto." *Wikipedia*, 1 Feb. 2023, en.wikipedia.org/wiki/Geography_of_Pluto.

4. Learning to See

1. The two quotes are from page 8 of Galilei, Galileo. *Sidereus Nuncius, or The Sidereal Messenger.* Translated by Albert van Helden, University of Chicago Press, 2016.

2. My source for Harriot is primarily Bloom, Terrie F. "Borrowed Perceptions: Harriot's Maps of the Moon." *Journal for the History of Astronomy* 9.2, 1978, 117-122.

3. My source for this is Edgerton, Samuel Y. *The Heritage of Giotto's Geometry: Art and Science on the Eve of the Scientific Revolution.* Cornell University Press, 1991. In particular, I am drawing upon the chapter reprinted as "Galileo and the Geometrization of Astronomical Space," in Danielson, Dennis R. *The Book of the Cosmos: Imagining the Universe from Heraclitus to Hawking.* New York: Basic Books, 2000.

4. See, for example, Bredekamp, Horst. "Gazing Hands and Blind Spots: Galileo as Draftsman." *Science in Context* 13.3-4, 2000, 423-462.

5. I list here the writings that I have found most useful, and which influenced my thinking in this essay: Daston, Lorraine

and Peter Galison. *Objectivity.* Princeton University Press, 2021; Daston, Lorraine and Elizabeth Lunbeck, eds. *Histories of Scientific Observation.* University of Chicago Press, 2011. The importance of idealization and the *International Cloud Atlas* are discussed in Daston, Lorraine. "On Scientific Observation," *Isis,* 99.1, 2008, 97-110.

6. The most recent version, updated by the World Meteorological Association, can be found online: *International Cloud Atlas.* cloudatlas.wmo.int/home.html.

7. Bruner, Jerome S., and Leo Postman. "On the Perception of Incongruity: A Paradigm." *Journal of Personality* 18.2, 1949, 206-223; cited in Kuhn, Thomas S. *The Structure of Scientific Revolutions.* University of Chicago Press, 2012, p. 63.

8. There is a huge literature on this topic. The main point I want to emphasize is the key idea that our simple and unitary experience of vision hides the fact that it is complex and distributed. See, for example: Marr, David. *Vision: A Computational Investigation into the Human Representation and Processing of Visual Information.* San Francisco: WH Freeman and Company, 1982. Marr notes several talks and papers in the 1970s by Elizabeth Warrington as an important influence on his thinking about the distributed nature of vision, and categorization. She, in turn, says that Marr heavily influenced her subsequent thinking about the computational aspects of vision. See, for example, Warrington, Elizabeth K., et al. "Warrington and Taylor's 1978 Paper." *Perception* 38.6, 2009, 933-947, which also includes a number of commentaries on her work by leading researchers. See, also, the sketches in the chapter by Farah, Martha J., and Todd E. Feinberg. "Visual object Agnosia." in *Patient-based Approaches to Cognitive Neuroscience,* 2000, 117-122.

9. "Lambda-CDM Model." *Wikipedia,* 30 Jan. 2023, en.wikipedia.org/wiki/Lambda-CDM_model.

10. A wonderful and very readable summary of LIGO's pattern matching strategy was outlined a generation ago by Kip Thorne in Chapter 10 of *Black Holes & Time Warps: Einstein's Outrageous Legacy*, WW Norton & Company, 1995. Thorne was one of the three co-winners of the 2017 Nobel Prize for his pioneering work on LIGO. An example of the 'chirp' that is the gravity-wave signal pattern for late-stage colliding black holes can be found in Wall, Mike. "Chirp! Here's What the New Gravitational-Wave Signal Sounds Like," *Space.com*, 16 June 2016. More background about how LIGO works can be found in Muller, Derek. "The Absurdity of Detecting Gravitational Waves." *YouTube*, uploaded by Veritaseum, 5 Jan. 2017, www.youtube.com/watch?v=iphcyNWFD10.

11. Sheehan, William. *Planets & Perception: Telescopic Views and Interpretations, 1609-1909*. University of Arizona Press, 1988.

12. Dreifus, Claudia. "In 'Half Earth,' E. O. Wilson Calls for a Grand Retreat." *The New York Times*, 29 Feb. 2016.

13. Dyson, Freeman J. *Infinite in All Directions: Gifford Lectures Given at Aberdeen, Scotland, April-November 1985*, Harper and Row, 1988.

5. The Weightlessness of Knowledge

1. Becker, Barbara J. *Unravelling Starlight: William and Margaret Huggins and the Rise of the New Astronomy*. Cambridge University Press, 2011.

2. See Forbes, Duncan A. "So you want to be a professional astronomer!" *Mercury*, Vol. 37, no. 2, Spring 2008, pp. 24-28. Forbes estimates the total number of astronomers is approximately 10,000.

3. Cambridge University Commission. *Report of HM Commissioners Appointed to Enquire into the State and Revenues of*

the University and Colleges of Cambridge: Together with the Evidence, and an Appendix. Cambridge University Press, 1852. (Reprinted July 2009.) The full text of the 1852 original is available online as a Google eBook. The relevant passages concerning the Plumian Professorship begin on page 59 in the original.

4. Crowther, James Gerald. *The Cavendish Laboratory, 1874-1974*. London: Macmillan, 1974, page 7-9.

5. Here's the full passage, (translation by A. S. Kline.): "Who, if I cried out, would hear me among the Angelic Orders? And even if one were to suddenly take me to its heart, I would vanish into its stronger existence. For beauty is nothing but the beginning of terror, that we are still able to bear, and we revere it so, because it calmly disdains to destroy us. Every Angel is terror." *Rilke, Rainer Maria (1875–1926) - Duino Elegies.* www. poetryintranslation.com/PITBR/German/Rilke.php.

6. Pilkington, Ed. "US Nearly Detonated Atomic Bomb Over North Carolina: Secret Document." *The Guardian*, 20 Sept. 2013, www.theguardian.com/world/2013/sep/20/usaf-atomic-bomb-north-carolina-1961. Accessed 8 Feb. 2023.

6. How Much Can We Afford to Forget, if We Teach Our Machines to Remember?

1. Pietrzak, Barbara, et al. "Education for the Future." *Science.* 360.6396, 2018, 1409-1412.

2. Wegner, Daniel M. "Transactive Memory: A Contemporary Analysis of the Group Mind." in *Theories of Group Behavior*, 1987, 185-208.

3. Sparrow, Betsy, Jenny Liu, and Daniel M. Wegner. "Google Effects on Memory: Cognitive Consequences of Having Information at Our Fingertips." *Science.* 333.6043, 2011, 776-778.

4. See, for example, CPG Grey's entertaining explanation: CGP Grey. "How Machines Learn." *YouTube*, 18 Dec. 2017, www.youtube.com/watch?v=R9OHn5ZF4Uo. Accessed February 9, 2019.

5. Buolamwini, Joy. "Opinion | When the Robot Doesn't See Dark Skin." *The New York Times*, 22 June 2018, www.nytimes.com/2018/06/21/opinion/facial-analysis-technology-bias.html.

6. Fisher, Matthew, Mariel K. Goddu, and Frank C. Keil. "Searching for Explanations: How the Internet Inflates Estimates of Internal Knowledge." *Journal of Experimental Psychology: General* 144.3, 2015, 674; Ward, Adrian F. "Supernormal: How the Internet is Changing Our Memories and Our Minds." *Psychological Inquiry* 24.4, 2013, 341-348. See page 344.

7. Clark, Andy, and David Chalmers. "The Extended Mind." *Analysis* 58.1, 1998, 7-19.

8. See, for example, Lebedev, Mikhail A, and Miguel AL Nicolelis. "Brain-machine Interfaces: From Basic Science to Neuroprostheses and Neurorehabilitation." *Physiological Reviews* 97.2, 2017, 767-837.

9. The science fiction writer William Gibson noted this shift in our interior perception of the exterior world in Gibson, William. "Opinion | Google's Earth." *The New York Times*, 1 Sept. 2010, www.nytimes.com/2010/09/01/opinion/01gibson.html.

10. *Google Search Statistics - Internet Live Stats.* www.internetlivestats.com/google-search-statistics. Accessed February 8, 2023.

11. "Advanced Chess." *Wikipedia*, 21 June 2022. "Advanced chess" is a term coined by former chess world champion Gary Kasparov, who has promoted hybrid man-machine play since his loss to IBM's Deep Blue computer in 1997.

12. Metz, Cade. "AI Shows Promise Assisting Physicians." *The New*

York Times, 12 Feb. 2019, https://www.nytimes.com/2019/02/11/
health/artificial-intelligence-medical-diagnosis.html.

7. The Yowl! In the Machine

1. See, for example, Volokh, Eugene, and Donald M. Falk. "First Amendment Protection for Search Engine Search Results—White Paper Commissioned by Google." *SSRN Electronic Journal*, Elsevier BV, 2012, https://doi.org/10.2139/ssrn.2055364.
2. Kurki, Visa. *A Theory of Legal Personhood (Oxford Legal Philosophy)*. Oxford UP, 2019.
3. Bayern, Shawn. "The implications of modern business–entity law for the regulation of autonomous systems." *European Journal of Risk Regulation* 7.2, 2016, 297-309.
4. Rossen, Jake. "The Tragic Life of Clippy, the World's Most Hated Virtual Assistant." *Mental Floss*, 28 Sept. 2017, www.mentalfloss.com/article/504767/tragic-life-clippy-worlds-most-hated-virtual-assistant.
5. Picard, Rosalind. "Rosalind Picard: Affective Computing, Emotion, Privacy, and Health | Artificial Intelligence Podcast." *MIT Media Lab*, www.media.mit.edu/articles/rosalind-picard-affective-computing-emotion-privacy-and-health-artificial-intelligence-podcast.
6. Miner, Adam S., et al. "Smartphone-Based Conversational Agents and Responses to Questions About Mental Health, Interpersonal Violence, and Physical Health." *JAMA Internal Medicine*, vol. 176, no. 5, American Medical Association (AMA), May 2016, p. 619. https://doi.org/10.1001/jamaintern- med.2016.0400.

8. We Are All the Tin Woodman

1. Wittgenstein, Ludwig. "A lecture on ethics." *The Philosophical Review*, 74.1, 1965, 3-12. [Quoted passages are at the very end.] Based on a lecture given at Cambridge in 1929 or 1930.
2. Dyson, Freeman J. *Infinite in All Directions: Gifford Lectures Given at Aberdeen, Scotland, April-November 1985*, Harper and Row, 1988, pps. 7-8.
3. Atwood, Margaret. "Am I a Bad Feminist?" *The Globe and Mail*, 9 July 2020, www.theglobeandmail.com/opinion/am-i-a-bad-feminist/article37591823.

9. Stirring the Ashes of our Dreams

1. JOY of MUSEUMS. "Missing Heads of the Kings From Notre-Dame De Paris." *Joy of Museums Virtual Tours*, 15 Nov. 2020, joyofmuseums.com/museums/europe/france-museums/paris-museums/musee-national-du-moyen-age/the-gallery-of-kings-heads.

10. Beware the Orwellian Trap

1. In the essays 'Learning to see' and 'Gaia shrugged', I examine this creative process in the context of discovery in the sciences.
2. See, for example, Woolgar, Chris. "The Medieval Senses Were Transmitters as Much as Receivers | Aeon Ideas." *Aeon*, 8 Feb. 2023, aeon.co/ideas/the-medieval-senses-were-transmitters-as-much-as-receivers.
3. Winer, Gerald A., and Jane E. Cottrell. "Does Anything Leave the Eye When We See? Extramission Beliefs of Children and

Adults." *Current Directions in Psychological Science* 5.5, 1996, 137-142.

4. See, for example, @RadioFreeTom. "These are people - again, especially the men - trapped in the eternal drama of adolescence. They are creatures of a leisure society, bored by the ordinariness of life, angry that the world is not more interesting and that others refuse to pay them their heroic due." *Twitter*, 14 Jan. 2021, 7:00 p.m., twitter.com/radiofreetom/status/1349869121983438851.

5. Cron, Lisa. *Wired for Story: The Writer's Guide to Using Brain Science to Hook Readers From the Very First Sentence*. Clarkson Potter/Ten Speed, 2012.

6. Ward, Charlotte, and Voas, David. "The Emergence of Conspirituality." *Journal of Contemporary Religion* 26.1, 2011, 103-121.

7. This cosmology of conspiracies is playfully examined in Eco, Umberto, and William Weaver (transl.). *Foucault's Pendulum*. First, Mariner Books, 2007.

8. Kahan, Dan M. "Climate Science Communication and the Measurement Problem." *Political Psychology* 36, 2015, 1-43.

11. Gaia Shrugged

1. See, for example, "Louisiana's Comprehensive Master Plan for a Sustainable Coast." June, 2017. http://coastal.la.gov/wp-content/uploads/2017/04/2017-Coastal-Master-Plan-Web-Book_CFinal-with-Effective- Date-06092017.pdf

2. Daston, Lorraine and Peter Galison. *Objectivity*. Zone Books, 2010.

3. *Sea Level Trends - NOAA Tides and Currents*. tidesandcurrents.noaa.gov/sltrends/sltrends.html.

4. See, for example, Key Message 4 of Chapter 2 of the most

recent US National Climate Assessment: *Climate Science Special Report Fourth National Climate Assessment (NCA4)*, Vol. 1. https://nca2018.globalchange. gov/chapter/2/.

5. "Homepage." *National Snow and Ice Data Center*, 7 Feb. 2023, nsidc.org/home.

12. Can We Create a Climate for Hope?

1. Moskvitch, Katia. "Rivers Under the Sea: Earth's Vital Waterways Are Also the Strangest." *New Scientist* 221.2957, 2014, 42-45.

2. See, for example, Ryan, William, and Walter Pitman. *Noah's Flood: The New Scientific Discoveries About the Event That Changed History.* Simon and Schuster, 2000, pps. 61-66.

3. Gardiner, Stephen M. *A Perfect Moral Storm: The Ethical Tragedy of Climate Change.* Oxford University Press, 2011.

4. Greenfield, Patrick. "'Sweet City': the Costa Rica Suburb That Gave Citizenship to Bees, Plants and Trees." *The Guardian*, 29 Apr. 2020.

13. Hooked on a Feeling

1. Heschel, Susannah (ed.). *Abraham Joshua Heschel: Essential Writings*, Orbis Books, 2017. The quote is taken from the essay "What Manner of Man is the Prophet?" available online at https://www.jewishideas.org/article/selected-writings-abraham-joshua-heschel.

2. See, for example, Kahneman, Daniel. *Thinking, Fast and Slow.* Farrar, Straus and Giroux, 2013.

3. The full title of Jamieson's book gives a sense of the tone: Jamieson, Dale. *Reason in a Dark Time: Why the Struggle for*

Climate Change Failed—and What It Means for Our Future. Oxford, United Kingdom, Oxford UP, 2017.

4. Robinson, Kim Stanley. "The Coronavirus is Rewriting Our Imaginations." *The New Yorker,* 1 May 2020, www.newyorker. com/culture/annals-of-inquiry/the-coronavirus-and-our-future.

5. Bellis, Mark A., et al. "Associations Between Adverse Childhood Experiences, Attitudes Towards COVID-19 Restrictions and Vaccine Hesitancy: A Cross-sectional Study." *BMJ Open* 2022; 12: e053915. doi: 10.1136/bmjopen-2021-053915

6. Professor Shore's lecture is #20 of the online course "Timothy Snyder: The Making of Modern Ukraine." *YouTube,* www.youtube.com/playlist?list=PLh9mgdi4rNewfxO7LhBo-z_1Mx1MaO6sw_.

15. A Return Volley, Forty Years Late

1. The play can still be viewed through *National Theatre at Home | National Theatre.* www.nationaltheatre.org.uk/national-theatre-online/. (Accessed 9 Feb. 2023.)

16. All Our Plans Will Go Awry

1. Stevenson, Robert Louis. "Reflections and Remarks on Human Life." *The Project Gutenberg eBook of the Works of Robert Louis Stevenson Volume XVI, by Robert Louis Stevenson,* p. 363. www.gutenberg.org/files/30990/30990-h/30990-h.htm.

2. Solnit, Rebecca. "Woolf's Darkness: Embrace the Inexplicable." *The New Yorker,* 24 Apr. 2014. www.newyorker.com/books/page-turner/woolfs-darkness-embracing-the-inexplicable. Woolf's journal entry was dated January 18, 1915. Solnit relates that Woolf had just recovered from a bout of

depression and was dealing like everyone else with the growing catastrophe that was the First World War.

3. This is a point often made by writer Annalee Newitz. See, for example, Alptraum, Lux. "Writer Annalee Newitz on the Perils of Historical Amnesia." *Medium*, 12 Dec. 2021, onezero. medium.com/writer-annalee-newitz-on-the-perils-of-historical-amnesia-4138c88958e6.

4. When I refused to sign the deed, I was assured that the Fair Housing Act of 1968 made all such covenants null and void.

5. In December 2022, the National Ignition Facility at Lawrence Livermore Laboratory in California announced that for the first time in history, a laboratory fusion experiment had exceeded 'scientific breakeven' in a laser-driven implosion. Without going into too many technical details, this means that the fusion energy output was greater than that of the laser beams driving the implosion. This did *not* take into account the conversion efficiency of the lasers, and there was no attempt to convert the output energy into electricity for the grid. This experimental result is therefore still very far from a commercial fusion reactor, but it is scientifically significant all the same.

6. For a summary, see: Ball, Philip. "The Chase for Fusion Energy." *Nature*, 17 Nov. 2021.

17. A Science Without Time

1. Augustine, St., and Henry Chadwick (transl.). *Confessions.* Oxford University Press, 1991, p. 230.

2. Ahmed, Issam. "What the World's Most Accurate Clock Can Tell Us About Earth and the Cosmos." *PhysOrg*, 9 Sept. 2021, phys. org/news/2021-09- world-accurate-clock-earth-cosmos.html.

3. See, for example, Carroll, Sean. "The Reality of Time." *Sean

Carroll Blog, 3 Apr. 2015, www.preposterousuniverse.com/blog/2015/04/03/the-reality-of- time.

4. My understanding here has been strongly influenced by the neuroscientist William Calvin, who in a series of books like *The Ascent of Mind* uses the metaphor of the 'Darwin machine' to describe a similar idea. Also, David Marr's *Vision;* Stephen Pinker's books about language; and Eric Kandel's *In Search of Memory.*

5. Kadare, Ismail, and Barbara Bray (trans.). *The Palace of Dreams.* Arcade Publishing, 1993.

6. Mlodinow, Leonard, and Todd A. Brun. "Relation Between the Psychological and Thermodynamic Arrows of Time." *Physical Review E* 89.5, 2014, 052102.

7. Feynman, Richard Phillips, and Laurie M. Brown. *Feynman's Thesis: A New Approach to Quantum Theory.* World Scientific, 2005.

8. Barbour, Julian. *The End of Time: The Next Revolution in Physics.* Oxford University Press, 2001.

9. The videos can be viewed online here: Tracy, Gene. "Why Doesn't Physics Help Us to Understand the Flow of Time?" *Aeon*, 9 Feb. 2023, aeon.co/es- says/why-doesn-t-physics-help-us-to-understand-the-flow-of-time.

Epilogue

1. Humphrey, Omar J., and Robert Rexdale. *Wreck of the Rainier: A Sailor's Narrative.* Franklin Classics, 2018. Also available online through Google Books, books.google.mg/books?id=-J7RCAAAAYAAJ.

2. The stories the Marshallese tell of these events can be found in Tobin, Jack A. *Stories From the Marshall Islands: Bwebwenato Jan Aelon Kein.* Amsterdam, Netherlands, Amsterdam UP,

2001. In particular, see "86. Story About an American Ship,"
p. 363; and "87. Story About an American Ship: Further Explanation," p. 366.